Poisons and Poisonings
Death by Stealth

Poisons and Poisonings
Death by Stealth

Tony Hargreaves

Print ISBN: 978-1-78262-717-3

A catalogue record for this book is available from the British Library

Published by The Royal Society of Chemistry,
Thomas Graham House, Science Park, Milton Road,
Cambridge CB4 0WF, UK

Registered Charity Number 207890

Visit our website at www.rsc.org/books

Printed in the United Kingdom by CPI Group (UK) Ltd, Croydon, CR0 4YY, UK

Preface

This book is written in the style of popular science rather than that of an academic text. It covers poisons, poisoning and poisoners. Beginning with the history of poisons and mankind's involvement with them, it then considers examples of poisons and outlines cases in which poisons were developed as weapons of murder and warfare. It continues with a discussion of how poisons cause injury or death in the human body. From here it looks at examples of poisons and their different origins, animal vegetable or mineral. Examples from these groups provide the basis for the next section in which comparisons are made between poisons and medicines.

Before the era of industrial chemistry, most poisons were of natural origin but with the manufacture of chemicals, poisoning began on a large scale. This formed the basis of chemical weapons used with horrific results in World War I where the intentional poisoning of humans was carried out on an industrial scale. Intentional poisoning for destroying human life and sanctioned by law is also considered in examples such as euthanasia, execution and abortion.

The book continues by taking a look at the methods available to the forensic scientist in search of evidence that puts the poisoner before the court and provides the evidence to enable justice to be done. In this section the procedures that relate to detection of poisons in dead bodies and exhumed remains are

Poisons and Poisonings: Death by Stealth
By Tony Hargreaves

Published by the Royal Society of Chemistry, www.rsc.org

discussed. This includes some detail about the chemistry and mechanisms in the process of death and decay.

Throughout the book case studies are reported in which the human element of poisoning is presented. This includes details of sentences handed out (life sentences and executions) to some of the more infamous poisoners.

To enable the reader to delve deeper, there is a list of poisoners and a glossary in which more technical details of poisons are given. As such, the book will prove useful to those with an interest in true crime, forensic science, industrial poisons, environmental poisons and food toxins, as well as the reader who has a general interest in the subject.

In writing this book I offer my thanks to the following. These include the University of Huddersfield library and academic staff, especially Gary Midgley and Tim Amsdon and students on the foundation chemistry course in the university's international study entre. I also wish to thank the following who have made helpful comments at the various stages of writing this book: John Cunneen, Jeffrey Simmons, Vivienne Ball and the staff at Jeremy Mills Publishing.

Introduction

Archaeological evidence in the form of stone tools and weapons shows us that early man was a creative thinker. He became skilled at flint knapping to produce knives, spears and finely crafted arrow heads. Some of those arrow heads were mounted on the shaft and held firm by means of a blend of resin and beeswax, strong enough to hold the flint in place as it penetrated an animal's tough skin.

There is little doubt that he was exploring materials and developing the skills to use them to improve his hunting. Making the resin and beeswax composite meant that he was carrying out trial-and-error experiments to optimise the physical properties of the mixture. With his ability to perform 'what if' experiments and his growing knowledge of plant materials, it is highly likely that he also experimented with poisonous substances from plants. Maybe this is when the poison arrow was invented, which is still in use in modern times by hunters in some parts of the World.

Mankind's creative thinking eventually moved him on from the Stone Age to the Age of Agriculture. No longer did he have to follow herds of animals in the hunt, for he had them in enclosures, enabling a kill to be made with relative ease. And at the same time he began cultivating fields of food crops. The Stone Age was giving way to the Age of Agriculture, which then spread Worldwide. People stopped being nomads and began to live in settled communities, permanently occupying an area of land.

Poisons and Poisonings: Death by Stealth
By Tony Hargreaves

Published by the Royal Society of Chemistry, www.rsc.org

Mankind was then faced with a new and serious problem. Whilst enjoying a reliable food supply, he also had to suffer the problem of pestilence and disease. You don't get something for nothing. His new problem had its roots in food storage and in growing groups of people living in crowded conditions. The disposal of waste was also something new, and brought more problems. Infectious diseases caused havoc, and plagues of insects caused panic. In his earlier days as a nomad living in small isolated groups, an infectious disease would not spread far. Furthermore, his nomadic lifestyle did not enable him to store large quantities of food, which are always a haven for insect pests, rodents and fungi.

Despite the problems, mankind battled on and used his creativity to seek a solution. He needed poisons to kill the rats, and he needed other poisons to use as medicines against the infectious diseases. Civilised living had begun and poisoning began in earnest. And so it goes on to the present time. We are still bound into the cycle of inventing a solution to a problem that was caused by the previous invention.

Today there is an extensive knowledge of poisons, how they work and where they come from (Figure 1). We may be poisoned by the food we eat, by the air we breathe, by the water we drink or by the medicines we take. Most of us have experienced poisoning that has resulted in a bout of sickness, and some of us have may have been poisoned more seriously and succumbed to long-term illness, or even taken to death's door.

The poison that has attacked us may have come our way accidentally through our occupations, or from use of harmful recreational substances, by ingesting microbes in a dish of dodgy curry from that back-street takeaway, or through disruption of our cells by a visiting virus.

However, the worst scenario for us is when a fellow human being is subjected to poison by stealth with the intent being murder. Administering poison with malicious intent is a serious crime in every civilised society. We find poisoning written about in the earliest of accounts that record murders by sneaky means. Poisoning continues in modern times. In the past 100 years, civilisation has experienced poisoning on such a vast scale that it would never have been imagined in earlier days.

Figure 1 With the skull and cross bones displayed on the jar of a chemical, there is no doubt as to what the risk of exposure is. This symbol is now accepted Worldwide in the labelling of toxic substances. © Shutterstock

Much of the content of this book is about the deliberate poisoning of one person by another individual or by several. Despite murder being the main focus, the other aspects of poisoning must be considered as part of the whole picture. Most of us have heard of arsenic, cyanide and strychnine. We know they are poisonous chemicals that have been widely used as murder weapons. In the literature on true crime, and in crime fiction, these three poisons have certainly made their mark. They are the classic trio and the stock in trade of many a crime writer.

Nowadays we hear little of them and regard them as confined to crimes of the past. But we must not think that poisons in general are confined to the past. Murder by poison is very much alive today, and preparing its victims for death tomorrow. We see new poisoning cases appear every few weeks in national news, and on an international scale every day.

It was not so long ago that Britain experienced the World's biggest serial poisoning carried out by one man. Although the exact number of victims will never be known, the evidence, up to the present, shows that he carried out around 300 murders. Dr Harold Shipman (Figure 2) was the killer. Over a period of years he chose his victims from his list of patients, and for his efforts he was given 15 life sentences. Only a few years before this, Beverley Allitt, a paediatric nurse, was convicted and given 13 life sentences for killing children in her care by administering poison.

In the chapters that follow, we examine poisoning from various angles. In learning about poisoning, it is essential to consider the science behind it. There are many books that explain the technical aspects of the subject, but they are for the specialist forensic scientist or toxicologist. Similarly, there are books that focus on the people involved, the poisoners and their victims. In this book, with its popular science approach, you will see a broader picture of poisoning and some of the forensic aspects of crime investigation.

Figure 2 The World's biggest serial poisoning was carried out by Dr Harold Shipman. Although the exact number of victims will never be known, it is estimated that he killed around 300 of his trusting patients by giving them lethal injections of diamorphine. He was given 15 life sentences.
Image courtesy of www.gutsandgore.com.

Science commands a good deal of respect when it comes to establishing the true facts. This is especially so in the investigation of serious crimes where forensic science generally plays a crucial role. In examining a poisoning incident, the investigator may believe, after looking at the circumstances, that a particular poison had been used. We must distinguish between believing and knowing. To understand the distinction is crucial.

Believing is subjective. It is based on gut feelings, intuition and instinct. We all have these, and they serve a purpose in protecting us from dangerous situations. If, for example, we encounter a man wielding a knife, our instinct is to keep well away from him. We believe he is likely to do us harm. Belief can often be a source of comfort, and this is at the heart of many belief systems such as the various religions. It has its uses, but to provide proof of an event we need more. We need knowledge and not belief. If a person is to be judged and possibly sent to the gallows, the circumstances must be known and the proof must be provided.

Science is the instrument that enables us to achieve this, and does so through a process of objective analysis of physical evidence. Today we take for granted the application of science to the investigation of crime. In Britain, as in many other industrialised societies, we are confident that the investigation of a suspected poisoning will be performed with thoroughness. We accept the competence of the police, the forensic scientists and the other professionals who are called upon to cast an expert eye over the evidence. Furthermore, the system is highly professional and is transparent enough to enable it to be monitored and regulated.

Nowadays the scientific method is accepted in the analysis of physical evidence from an incident scene. In the past, before the age of science, things were different. In the Middle Ages in Britain there were occasions when people were lawfully executed on the basis of witness testimony alone. The only evidence on offer came from a patchwork of verbal accounts from witnesses who may not have been the most reliable, or who wanted to influence the outcome. In some cases, there were forced confessions.

The witch hunts serve as an example of just how far belief can be taken. A woman being old, warty, owning a black cat and having a besom broom by the back door must be skilled in the art of sorcery. And if she lived within a two-mile radius of a suspicious death, this was sufficient cause for her to be accused of casting a spell by witchcraft. The death penalty was imposed, and she was executed by being burnt at the stake. In those days there was total reliance upon what people described. As we now know, this is unreliable even when the witness can be trusted to tell the truth. It is a fact that, when asked to recall the details of a particular sequence of events, each person will offer a different version of what took place – and when and where and by whom.

No matter how hard we try to be objective and deal with the true facts, we often find that we are subjective and allow belief to creep in. We form our own ideas and we develop our own theories. In science we must be aware of this, and keep our thinking clear and disciplined.

To enable us to organise our thinking on poisons we need to learn some facts in general, then we can go on to examine it in depth. Hopefully, we then discover the truth. For example: what at first sight appears to be criminal, turns out to be accidental; what looks like murder, is found to be natural causes; and what seems to be suicide, is shown to be murder. In learning about poisons we soon appreciate that our study is extensive. But our study of poison must also be intensive as it requires us to understand and explain.

We begin with a brief history of poisons and poisoning in *Primitive Potions and Poisons*. In this we will trace back the origins of poisons, learn of some historical cases, and see how criminals actually developed poisoning into a business and carried out poisonings on behalf of clients. In looking into the history we will note the origins of different poisons and the discovery of new examples.

From the history we go on to look at what actually happens when poisonous chemicals enter the human body, in *Chemical Chaos*. For a poison to act upon the human body it must first make its way into the body. We examine various modes of entry and how each enables the poison to be distributed through the body until it reaches a particular organ, wherein it sets about damaging the tissue with fatal consequences. To understand a poison's destructive action, we see some of the chemical processes that form the basis of poisoning mechanisms.

In *Animal and Vegetable*, most of the focus is on poisons found in plants, because this covers a huge range. It is here that we come across chemicals that are used as medicines to cure, and, in larger doses, as poisons to kill. Some of the examples go back many centuries but are still used in medicine today – and in poisoning today. Here we see examples of modern synthetic drugs, and how, instead of saving lives, they are used to destroy lives. In one case study we consider some aspects of Marilyn Monroe's death that demonstrate how difficult it can be to reveal the truth, the whole truth and nothing but the truth.

The chapter titled *Mineral Matters* examines materials on the Earth, in the Earth, and arriving on the Earth from space. A few metals are highly poisonous, as is seen when we consider different mineral sources. Here we come across recent examples of how, due to man's neglect, metals have caused havoc in the form of mass poisoning. We might be surprised to find that poisonous chemicals based upon lead were used by the ancient Romans to sweeten their wine, and that solutions of arsenic were sold as tonics up until the middle of the 20th Century.

Poison or Medicine is the theme in the following chapter. All medicines are poisons. We need only look at the dosage details and the warnings of overdose that accompany the medicines we take. If we take too little, it will have no effect; take the right amount and our suffering may be relieved; take too much and death results. And what is a lethal overdose for one person may have no effect on another.

Many poisons are in fact medicines as well. Take arsenic, for instance. It is the best-known of all poisons and has destroyed lives over many centuries since its purification back in the days of Arab alchemists and their alembics. But for many years arsenic was a popular tonic; a pick-me-up rather than a put-me-down. It was widely used by men with poor penis power, making it the Victorian version of Viagra. Arsenic played an important role when it was formulated into the antibiotic Salvarsan, which was the first successful cure for syphilis. Poisonous mercury was used up until the 1960s in Golden Eye Ointment for treating a sty on the eyelid. Mercury was also the active ingredient in teething powders for babies.

It is not surprising then that, when medicine is being administered, great care must be taken and procedures must be well defined. The case of Marilyn Monroe is relevant here. She was treated by two different doctors and with different medications, with one doctor not knowing what the other was doing.

It was in the 19th Century that industrial chemicals were developed, and with them came new opportunities for the poisoner. The contribution made by industrial chemicals is considered in *Man-made Menace*. We also note that this marked the onset of large-scale environmental poisoning, the effects of which are still increasing today. Poisonous industrial chemicals cause havoc when they inadvertently enter the environment, and some of this amounts to criminal neglect. We note a recent example of how one poisonous industrial chemical was intentionally added to commercial baby milk with horrendous consequences. The consequences for the culprits were also horrendous, for they ended up being executed.

The next chapter, *From Poison to Prison*, introduces us to the forensic science of poisoning. It is through forensic analysis of physical evidence that we are able to identify poisons, and bring to justice those responsible. In this chapter we see some of the early forensic tests that required the analyst to identify certain poisons by tasting body fluids and tissue from a dead body – and sometimes from an exhumed body with the flesh being in an advanced state of decomposition. From crude tests such as these, forensic science has come a long way, and now uses methods such as neutron activation analysis in which a tiny section of a human hair may be analysed for poisonous metals without the slightest damage to the sample.

We must learn something of the processes of death and decay to appreciate the problems of analysing a dead body. In some instances, the body has been removed from the crime scene and concealed, or efforts have been made to destroy it. We note the methods that have been employed by a killer to dispose of the evidence, such as crude burial or futile attempts at burning the corpse. Investigating a crime often also takes account of the psychology of the perpetrator. Here we take a look at the psychological profile of poisoners.

Gruesome but necessary examinations must often be performed in the search for clues to a suspicious death. In *Chemistry, Clues and Crime*, learning about how and when a person died involves carrying out detailed tests on their remains.

Here we take a look at forensic investigations by focussing upon case studies.

In considering the chances of poisoning being detected, it is worth noting that a dead body will contain residues of the particular poison. The victim takes the murder weapon with them to the grave, and so a poisoner can never be confident of remaining undetected. Bodies are often exhumed from makeshift graves and lawful graves even years after death. Of course, if the body has been cremated, and efficiently cremated, this destroys much of the evidence, but perhaps not all of it. Bones do not burn.

Unlike earlier times, when poisons could be purchased over the chemist's counter on the pretext of poisoning vermin, our modern safety regulations prevent this. Poisons have now been removed from general circulation, and poisonous formulations for domestic use have been replaced by safer alternatives. Furthermore, detailed records of poisonous chemicals are kept. There is traceability in the chain of supply, and amounts are recorded.

When a poison falls into the wrong hands, it has the potential for use as a murder weapon. However, anyone considering such an evil deed might also consider the armoury that is available to forensic science. Forensic science is now highly developed, can provide results quickly, and is readily available to all police forces. Furthermore, analysis can be carried out on minute samples that are too small to be seen by the human eye, and sometimes cannot even be seen under the laboratory microscope.

Clearly, in modern times, the chances of someone getting away with poisoning are extremely slim. It is worth noting that, on the one hand, the people who investigate poisonings are experienced professionals who know just what to look for. On the other hand, the person responsible for the poisoning is usually an inexperienced amateur who is doing it as a 'one off' and who will be noticeably ill-at-ease once some evidence comes to light and the law starts to tread on his toes.

Contents

Poisons and Poisonings: Death by Stealth
By Tony Hargreaves

Published by the Royal Society of Chemistry, www.rsc.org

CHAPTER 1

Primitive Potions and Poisons

It is London in the 1890s. A young woman has just taken some pills given to her by a doctor who said she was looking pale. About 10 minutes pass and she starts to experience breathing difficulties. Another minute and she collapses.

Her head and neck muscles go into a spasm. Following this, her facial muscles force the mouth into a hideous and exaggerated grin known as *risus sardonicus.* The whole of her face becomes liverish red. Another minute on sees her entire body taken over by convulsions as the spasms take over every muscle.

She lies there shaking violently. Suddenly her abdomen is forced upwards as her backbone arches, leaving only her head and heels touching the ground. Now she starts to slowly suffocate as her diaphragm becomes paralysed and stops her lungs working.

Mercifully, death arrives but the poison has not yet finished, for the process of rigor mortis has set in with unusual rapidity. Her body is frozen into a rigid and contorted mass. A most agonising death and a frightening sight to witness. This is the horror of strychnine, the nastiest of poisons. It tortures its victim before allowing death to rescue them from their hell.

Despite knowing all the horrors of this poison, Dr Thomas Neill Cream, later to be known as the 'Lambeth Poisoner', used it to kill four prostitutes. And who knows how many other victims

Poisons and Poisonings: Death by Stealth
By Tony Hargreaves

Published by the Royal Society of Chemistry, www.rsc.org

experienced the horror of strychnine, for it was by no means an uncommon poison.

Poisoning rises above all other means of murder in terms of our curiosity. We are fascinated by it. Perhaps this is because poison, in its role as the invisible weapon, is administered by stealth. There is something sinister about poisoning that invades our comfort zone.

1.1 ANCIENT EXPERIMENTS

When did humans discover that exposure of the body to certain substances results in death? In order to search for the answer, we must turn to archaeology. Some of the oldest archaeological finds are estimated to have been made half a million years ago in the Old Stone Age (the Palaeolithic). The earliest finds were made from nearby sources of random stone. Later, flint became widely used. Studies of flint weapons and tools such as arrow heads, spears, axes and scrapers show the versatility of this material, especially its property of breaking to form a razor-sharp edge.

Fabricating these items needed a good deal of skill in the art of flint knapping. We see from the evidence that there was much creativity. Furthermore, it was combined with systematic working. Clearly, ancient man was learning through a sequence of 'what if' experiments. What we refer to today as the 'scientific method' was at work in Stone-Age times, bringing about the technology to make certain tasks more efficient.

Fixing a flint arrowhead to a shaft was no easy task, for the flint must remain firmly attached when it hits the tough skin of the animal being hunted. The flint piece must penetrate the flesh and not fall off on impact. Chemical analysis of archaeological finds provides the evidence. The results show that a flexible resin was used to attach the arrowhead to the shaft. Furthermore, the resin was birch bark made flexible with beeswax. It had just the right composition to act as a strong adhesive capable of withstanding the mechanical forces imposed on the arrow.

It seems that experiments with resins and plasticizing substances were being performed, as such a material is not available as a natural substance. This is probably mankind's first experimental work in making a material with just the right properties. Ancient man, it seems, was trying his hand at chemical formulation.

It is at this stage that he may have discovered that certain substances extracted from plants were poisonous, and placing those substances on the arrowhead brought down an animal quickly. This would have reduced the time spent in chasing the animal while it slowly bled and collapsed. Poisoning had been discovered. The discovery had been made by means of experiments. The scientific method was at work: experiment; result; interpretation; new experiment and new result. And so on until the process was at its optimum.

Large numbers of flint weapons and tools have been found Worldwide and studied in detail with modern analytical instruments. In some countries people still use arrows and darts with a dab of sticky poison on the tip. For example, most of us know about the South American communities of hunters that use darts tipped with curare.

With early man's discovery of poisoning there must have been many who were keen to learn more of the new technology. They may have had some ideas of poisonous substances. If they ate certain berries, they would become ill and die; if they chewed a particular type of leaf, they would go crazy; if they tasted some types of plant juices, they would experience convulsions and painful paralysis. As their nomadic lifestyle took them into new environments, people would encounter different plants and learn that some must be avoided.

However, when looking at possibilities from pre-historic evidence, there is always an amount of conjecture – albeit intelligent conjecture. It was not until humans abandoned their hunter-gatherer lifestyles and took to living in settled groups in agricultural communities that the first written accounts of poisons appeared.

1.2 CLEOPATRA'S COBRA

One of the oldest known works to report poisons is the Ebers' Papyrus of 1550 BC. It gives recipes for medicines aimed at everyday illnesses, and formulations for a huge range of poisons to kill household pests. The formulations are based upon plant-derived poisons such as hemlock, mandrake, aconite, opium and the heavy metals arsenic, antimony and lead.

The Egyptians used the 'penalty of the peach' as part of their system of justice. The guilty were made to swallow the distillate from crushed peach kernels, an aqueous solution of a treacherously poisonous form of cyanide known as 'prussic acid'. There were other widely used means of poisoning. For example, on learning of the death of her lover Mark Antony, Cleopatra poisoned herself with the venom of an asp, after having had her servants (or criminals) try out different poisons upon themselves. There is some uncertainty as to the exact species of snake, but opinion comes out in favour of the Egyptian cobra.

Poisoning by the venom of an asp was used as a death penalty, but it was not applicable to everyone. It seems to have been reserved for more respected members of the community who, despite their particular crime, were regarded as deserving a dignified death. Apparently, the bite of an asp was the least torturous means of execution, the victim simply becoming drowsy and falling asleep before death claimed them – a far cry from the terror of some plant poisons like the strychnine we came across earlier.

In Act 5 of Shakespeare's tragedy *Antony and Cleopatra*, we find reference to her suicide.

With thy sharp teeth this knot intrinsicate
Of life at once untie: poor venomous fool
Be angry and despatch.

1.3 PYTHAGORAS' POTIONS

The ancient Greeks knew a lot about plant poisons and heavy metal poisons such as arsenic, antimony, mercury and lead. They also had a range of antidotes. A combination of this know-how and the ready availability of poisons explained why suicides and murders were so common. However, poisoning was not confined to the killing of individuals, but was also used for mass poisoning. Solon, the Athenian statesman, during the siege of Cirrha in about 590 BC, put poisonous hellebore root into an aqueduct from which the enemy drew water.

Criminals were executed with hemlock, as in the famous example of Socrates. According to one of his disciples, Plato, he was condemned in 399 BC at the age of 70. The penalty was that he must drink hemlock, a poison known as the 'Athenian state poison'. The charge against Socrates was heresy and corruption. Apparently he showed contempt for conventional ideas, and had been corrupting the youth with his theories. The mathematician Pythagoras (570–480 BC), when not pondering the square of the hypotenuse, spent time studying the poisonous effects of metals like tin, mercury, lead and copper upon the human body.

Another great name from the Greek archives was Hippocrates (460–377 BC), who studied the workings of the human body. He also was involved in poisons, especially from the point of view of how to deal with and treat poisonings by purging them from the body with enemas and emetics. In his writings he noted over 400 drugs, among which plant poisons such as those from henbane and mandrake were listed.

Nicander of Colophon (185–135 BC) had extensive knowledge of poisons, as is evident from his poems that refer to venomous animals, how they deliver their poisons, and the plant poisons henbane, hemlock, colchicum and aconite. In his study of therapeutic drugs, Dioscorides (40–90 AD) mentions mercury, arsenic and compounds of lead and copper that we now believe to be sugar of lead and copper oxide. He classified poisons as animal, vegetable or mineral, and it appears he was the first to describe the toxic effects of mercury.

The highly regarded physician Galen (129–216 AD) made a major contribution to the knowledge of poisons by pulling together all that was known on medicine. His work was so thorough that it remained an authoritative reference in medicine for 1400 years after its completion.

1.4 MITHRIDATES' MAD HONEY

The King of Pontus, Mithridates VI (132–63 BC), was concerned that his mother was intent upon killing him. His suspicion was not entirely without foundation, for she had assassinated her husband. Mithridates began taking small doses of poisons to build up his resistance to them. He eventually became a formidable enemy of Rome, but was always obsessed with the fear of being poisoned. Taking poisons became part of his daily routine as he continued his efforts to build up immunity. For example, he drank blood from the ducks that fed on poisonous plants. His theory was that, if the ducks could happily live on poisonous food, then they must have, within their blood, a means of resisting or destroying the poison.

His efforts at building immunity gave rise to the word 'mithridatize' which is still in use today. (The Concise Oxford Dictionary, mithridatize: render proof against a poison by administering gradually increasing doses of it.) He systematically studied poisons, testing them on criminals awaiting execution and evaluating the effectiveness of antidotes. One of his antidotes, antidotum mithridatum, was based upon herbal remedies. Containing 15 ingredients, it became something of a universal antidote. Confidence in it is shown by its use through the ages. In fact, it was still readily available in Italy up until the 17th Century.

The Roman general Pompey launched an attack upon Mithridates. As a part of Mithridates' response, he placed pots of 'mad honey' in positions where unsuspecting Roman troops would find them and eat the honey. Mad honey is from the nectar of a species of rhododendron that grows around the Black Sea, and which results in the honey being poisonous due to the presence of grayanotoxin.

Needless to say, after devouring the honey Pompey's soldiers became too ill to defend themselves and were wiped out by Mithridates' men. Eventually Mithridates was defeated by Pompey, upon which he tried to commit suicide using poison. He failed because of the immunity he had built up. We find an interesting reference to King Mithridates' immunity in a poem in the collection *A Shropshire Lad* by A. E. Housman:

There was a king reigned in the East:
There, when kings will sit to feast,
They get their full before they think
With poisoned meat and poisoned drink.
He gathered all that springs to birth
From the many-venomed earth;
First a little, thence to more,
He sampled all her killing store;
And easy, smiling, seasoned sound,
Sate the king when healths went round.
They put arsenic in his meat

And stared aghast to watch him eat;
They poured strychnine in his cup
And shook to see him drink it up:
They shook, they stared as white's their shirt:
Them it was their poison hurt.
- I tell the tale that I heard told.
Mithridates, he died old.

The practice of building up resistance to poisons still goes on today. For example, those who handle venomous animals take increasing doses of the respective venoms to build up their resistance. Tolerance of poisonous substances is experienced by many a drug addict, who needs to increase the dosage because his drugs seem to become less effective in producing the high. And when that drug fails to deliver, a new drug is experimented with. The same tolerance effect is also seen in the alcoholic.

1.5 VOLCANIC VOLATILES

Poisoning for suicide and murder was common, and is to be found in records of the Roman period going back to the 4th Century BC. Poisoning at the dinner table was frequently used as a means of assassination, and in 331 BC the extent of the practice was officially recorded. There were professional poisoners; for example, the infamous trio Canidia, Martina and Locusta. Locusta's poisoning talents were made use of by the emperor Nero to eliminate some human obstacles. The success of the arrangement led to Locusta being made adviser on poisons and the setting up of a state-approved school of poisoning. The idea was that she would teach others the art of poisoning and how to defend Nero against the poisonous ambitions of his adversaries.

The Roman author Pliny the Elder (23–79 AD) produced a major work known as *Natural History*, in which were preserved many of the early ideas on medicines, poisons and antidotes. His work was respected as an authority on scientific matters up until the Middle Ages. It is noteworthy that he regarded poisons as being useful to relieve someone of the burden of living when their life became unbearable. Suicide and euthanasia go back a long way.

Pliny's expertise in poisons was well known and resulted in him being sent to Mount Vesuvius to study the eruption and reassure the people. It is ironic that he died when overcome by fumes from the volcanic activity. In effect, he was poisoned by sulphur dioxide gas. Roman poisoning practices favoured the use of substances extracted from poisonous plants, with hemlock, belladonna, hellebore, colchicum and aconite playing a prominent role.

1.6 ARABIC ARSENIC

During the period 500 to 1450 AD, important developments were made by the Arab alchemists. In particular, they invented methods for extracting and purifying chemicals. For example, the techniques of distillation, sublimation and crystallisation were introduced, which enabled major progress in what later became the science of chemistry. Jabir ibn Hayyan (8th Century) produced a pure form of arsenic known as 'white arsenic'.

The white arsenic meant that, for the first time, arsenic was available as an odourless and tasteless poison. It was these very properties that led to it becoming the most widely administered chemical for intentional poisoning. Prior to this white arsenic, the minerals that were available were highly coloured sulphides of arsenic, orpiment and realgar. It was the orange–red realgar that was used as the raw material for making white arsenic, which is arsenic trioxide.

1.7 MONKS AND MANDRAKE

In Europe during the Middle Ages there were few texts written on poisons and poisoning, and most of what was in circulation was from previous times. This was the period of the monasteries, and most new knowledge came from, and was kept within, religious orders, with the monks preparing and selling herbal concoctions as medicines and tonics. One of the tonics, Benedictine, was based upon alcohol, and retains its popularity even to this day.

Texts on herbs and poisonous plants were written by the monks, but stayed within the monastic confines as most of the outside population were illiterate. However, there was one notable text that appeared in 1424. This was the *Book of Venoms* by Magister Santes de Ardoynis, in which reference was made to arsenic, aconite, mandrake and cantharides (Spanish fly).

Information on poisoning may well have been somewhat limited, but murder by traditional poisons continued. The emergence of the apothecary made many toxic chemicals available: medicines; drugs; tonics; antidotes; and poisons for killing household pests. Without doubt this worked to the poisoner's advantage, as we see in Chaucer's *Canterbury Tales* (*ca.* 1387). The Pardoner's Tale refers to the apothecary:

And forth he goes – no longer he would tarry –
Into the town unto a 'pothecary
And prayed him that he woulde sell
Some poison, that he might his rattes quell...
The 'pothecary answered: "And thou shalt have
A thing that, all so God my soule save,
In all this world there is no creature
That ate or drunk has of this confiture
Not but the montance of a corn of wheat
That he no shall his life anon forlete
Yea, starve (die) he shall, and that in lesse while
Than thou wilt go a pace but not a mile
The poison is so strong and violent.

To ward off the evil spirits and poisons, talismans and amulets were favoured by those people most at risk. Mary Queen of Scots carried toadstones with her. The business of making and selling these magical objects was largely down to certain Jews, who had a mixed reputation with respect to poisoning. On the one hand they sold the charms; on the other hand they carried out poisonings.

1.8 RENAISSANCE MEDICINE

During the Renaissance, Paracelsus (1493–1541) studied medicine, and received his doctorate in 1516. He regarded the human body as like a chemical laboratory, and said of poisons: "What is there that is not poison? All things are poison and nothing is without poison". From this it was possible to start thinking in terms of the dose–response relationship: from no effect to lethal. Later, the British toxicologist Alfred Swaine Taylor (1806–1880) stated: "A poison in a small dose is a medicine, but a medicine in a large dose is a poison". In this context it is worth noting that water ingested in large enough volumes is poisonous, despite the fact that the human body is itself about 70% water.

Perhaps the most infamous of all the poisoners were the Italians Cesare Borgia (1476–1507) and his sister Lucrezia (1480–1519). They murdered several people with their own toxic formulation 'La Canterella', which was based on arsenic with secret additives according to a recipe obtained from a Spanish monk.

In Venice, a group of alchemists came together, calling themselves 'The Council of Ten' with the aim of arranging poisonings for the state. Anyone could be disposed of by payment of the appropriate fee. The favoured poison recipes listed the following ingredients: corrosive sublimate, along with the oxide, chloride and sulphide of arsenic. In addition to providing a service to the state, the wealthier members of society were also clients. The well-to-do had a greater need, for they were often involved in removing someone who was an obstacle to an inheritance.

Poisoning in this period was not necessarily a secretive affair, as is evident from the schools for poisoners that were set up in Venice and Rome. Furthermore, *Magiae Naturalis*, written by Giovanni Battista Porta (1535–1615) includes references on the art of poisoning. The book has details of how to carry out poisoning, such as how to poison someone's wine. It also includes the recipe for a strong poison known as Veninum Lupinum that was based upon aconite, arsenic and other lethal ingredients.

Even the highly respected Leonardo da Vinci (1452–1519) involved himself in poisons. Performing experiments, he aimed to increase the effectiveness of poisons in warfare. One of his ideas was to throw a mixture of powdered chalk and arsenic at the enemy, a suggestion that does not quite befit a man of his calibre. There must have been many people thinking that he should return to the drawing board and get back to designing helicopters.

Testing poisons to assess how quickly they could kill someone was often performed on criminals. However, Catherine de Medici (1519–1589) took this one step further and tried her concoctions out on the poor and the sick. Her method of assessment was based upon symptoms, potency and site of action.

1.9 PARISIAN POISONERS

Renaissance Italy was clearly a hive of activity in terms of murder by poisoning, but the French were also to make their mark in deliberate poisoning. It was estimated that in Paris alone, during the 1570s, there were 30 000 people involved with poisonings. The French poisoner Marquise de Brinvilliers (1630–1676) poisoned her father, two brothers and her sister. Not content with this, she went on to poison her own daughter. She was eventually convicted, tortured and beheaded.

The killing of unwanted husbands has always been a popular use for poisons. Once the romance has gone out of the relationship and a wife sees her husband as a boring and obese sluggard who has nothing to offer, she is to be forgiven for considering likely escape routes. Recognising the business potential of this, Catherine Marvoisin (1638–1680) set up to sell poisons to wives, and was responsible for the deaths of thousands of husbands before the authorities caught up with her and had her burnt at the stake.

The commercial side of poisoning was also of interest to Guila Tofana (1635–1719) who developed her own formulation, Aqua Tofana. This she sold in little vials that had a picture of St Nicholas on the label. The concoction, which contained arsenic, was ostensibly for improving a woman's complexion by curing skin complaints. Used in a particular way, this may well have been the outcome, since arsenic has this curative effect. However, another way of using it was for murder, and verbal instructions were provided for this, with the effect that over 600 people, among them two popes, met their deaths. Madame Tofana was executed in Naples in 1719.

The spirit of scientific enquiry that had begun in the renaissance flourished, and the examination of poisons and their effects played a significant part. For example, the physician Felice Fontana built upon the ideas of Paracelsus by studying animal toxins such as snake venoms. He concluded that these were transported in the body to a target organ. Further work took him on to study plant poisons, in particular the laurel berry and curare. The quality of his work made him the finest of the Italian chemists of the 18th Century.

In Britain there was the publication in 1702 of *A Mechanical Account of Poisons* by Richard Mead of St Thomas' Hospital, who was physician to George I and II, Newton and Walpole. He studied venoms, opium and poisonous gases. He described how venom is not effective if swallowed, but must be administered by puncturing the skin. To make his point, he swallowed a dose of venom with no ill effect. We now know this was very risky, for if he had had any broken skin, such as an ulcer or a wobbly tooth, the venom could have entered his bloodstream. The outcome would have been very different.

The new world of science was showing the way, and what had been alchemy with its mishmash of unreasoned experiments, was being changed into chemistry by the likes of Joseph Priestley (1733–1804) and Antoine Lavoisier (1743–1794).

1.10 FATHER OF FORENSICS

In 1813 Mathieu Orfila published a major work on the scientific study of poisoning, and earned himself the title 'father of toxicology'. He classified poisons into groups such as: corrosive; astringent; acrid; narcotic; and putrefacient. The English translation appeared in 1816. Orfila's experimental work included testing the toxicity of strychnine that was put into many popular tonics and considered safe. In his earlier work on forensic toxicology he studied the effect of strychnine on dogs, but as his experience developed he went on to examine many deliberate attempts at poisoning.

In 1829 Robert Christison, professor of legal medicine at Edinburgh University, wrote his *Treatise on Poisons* and gained recognition as the first British toxicologist. As we shall see below, it was at one of Professor Christison's lectures that Dr George Henry Lamson learnt the rudiments of toxicology before going on to murder his brother-in-law.

The mid-19th Century saw poisoning come to the fore, and there was much fear, out of all proportion to the risk, of being poisoned. There were several reasons: poisons were not difficult to obtain; and the growth in literacy meant more people could read about poisoning cases in newspapers. For many people it appeared that there was an epidemic of poisoning, especially arsenic poisonings, in Britain, France and Germany.

In the case of Marie Lafarge, suspected of murdering her husband with arsenic, the body was exhumed for testing after doubt had been cast on the reliability of the original examination. The remains of the organs were analysed for arsenic using a test developed by James Marsh, and first described in a scientific journal in 1836. After some retesting, the result indicated arsenic, confirming the guilt of Marie Lafarge. Here we have the first trial in which evidence from forensic analysis was accepted in court.

Another father figure in the field of toxicology is Alexander Gettler, who worked in the toxicology laboratory in New York up until the late 1950s. He was the first forensic chemist to use a spectrometer to identify thallium as the poison used by the wife of a Frederick Gross to kill her four children. Gettler is known as the 'father of American toxicology', which begs the question: Is American toxicology different from toxicology in other countries?

1.11 REST ASSURED

The introduction of life assurance brought with it the motivation to murder, for now it was possible to put a price on a life. Poisonings increased from the 1820s to reach a peak at 1850. Once again, the use of poison as a means of murder had become popular. As in the previous centuries, many murders were connected with inheritance, and so it kept its reputation as a family affair.

Taking out a large policy on a wife and then dispatching her to an early grave was, for the unscrupulous, simply a business opportunity. Choose the right wife or husband and then move in for the killing. But to defraud the insurance companies, subtle means of murder were called for. Poisoning was the ideal instrument, with arsenic holding the prime position, but cyanide was also becoming significant. However, as the motivation to poison someone became greater, so did the means of detecting poisons, as forensic science was also growing.

In an attempt to control the sale of poison in Britain, the Arsenic Act of 1851 was brought in. This was to ensure that the vendor would sell only to those he knew, and that both he and the purchaser would sign a poison's register. The Act also recommended the addition of colourants such as soot or indigo so that the arsenic was easily recognisable and less likely to be mistaken for another chemical. However, the Act was only of limited effectiveness, as became evident in the Bradford poisoning case of 1858 when white arsenic found its way into sweets that poisoned many children.

1.12 INDUSTRIAL-SCALE KILLING

The early part of the 20th Century saw the introduction of poisoning on an industrial scale. At 5 pm on the 22nd April 1915 German troops at Ypres demonstrated that intentional poisoning need not be confined to a handful of victims but could kill and maim huge numbers. Allied troops of World War I were introduced to the horrors of chemical warfare with the release of 180 000 kg of chlorine gas intended to poison them. Later, other poisons were tested: phosgene; prussic acid; and the vesicant mustard gas. Germany's enthusiasm for mass poisoning was also demonstrated during World War II in the Holocaust, in which the poison was prussic acid gas.

With the threat of chemical warfare, other countries embarked upon the development of their own poisonous gases. Though many of these were never used, knowledge of their manufacturing details was well established. From time to time we see chemical weapons such as nerve gases become a threat. For example, mustard gas was used by Iraq against Iran in the 1980s, and in 1995 we saw the Tokyo subway attacked by warfare gas.

With forensic science, advanced medical technology and the effective legislation and control of chemicals, we might expect there to be less deliberate poisoning going on. We note that substances such as good old arsenic are no longer readily available, but there are new chemicals coming onto the market on an almost daily basis.

There are drugs, all of which are poisons when taken in excess of the prescribed dose. There are agricultural pesticides (insecticides, rodenticides, fungicides, molluscicides, herbicides and others) specifically designed for the purpose of poisoning living systems and that may include humans. And there is also a huge and growing range of industrial chemicals.

We now have a way of living that relies heavily upon synthetic materials and practices from industry. In order to profit from the economy of large-scale operations, industry handles materials and chemicals in huge amounts. Poisons never before known are now being manufactured, stored and transported in huge quantities. All too often we hear of disasters involving poisons from industrial activities; for instance: the Minamata Bay mercury poisoning; the Bhopal isocyanate poisoning; and the Seveso dioxin poisoning.

Although much of the focus in this book is on the poisons used for intentional killing, we need to take account of poisoning in general, because there will always be occasions when evidence, or lack of it, leaves a poisoning as a borderline case. To appreciate the whole scene, we must take account of what is happening in the background and what is standing in the wings waiting to make an appearance. With this as an aim, we must see some cases of industrial poisoning, and other cases where poisoning has resulted from our natural environment. Global warming and poisoning of the planet will likely lead to the death of millions, as it is a form of chronic Worldwide poisoning brought on by us neglecting our duty to care for our planet.

We might question why, with all the poisons now available, there are not more poisonings. No doubt there are some poisonings passing unnoticed, being written off as deaths due to natural causes. Unlike other means of murder, poisonings can keep their secrets, as there may be no outward signs to arouse suspicion. Poisoning is death by stealth.

Interesting recent additions to the history of intentional poisoning appear from time to time, and we shall examine some of the main ones in the chapters that follow. Some high-profile cases of the last few decades are: Bulgarian writer Georgi Markov being poisoned with ricin in 1978 for causing embarrassment to the state; Viktor Yushchenko (2004) being near death's door after suffering dioxin poisoning; and Russian dissident Alexandere Litvinenko (2006) dying from the radioactive poison polonium-210.

1.13 BABY KILLERS

In most poisoning cases the victims are adults. However, in 19th Century Victorian Britain, many people were in the grip of poverty and lived on near-starvation rations. Having young mouths to feed was a problem, but when new mouths came along, usually unplanned, they were treated as nothing more than a burden. Often steps were taken to kill the newly arrived offspring. But if the parents could barely afford to buy food, surely they did not have the means to purchase poison?

To put this into context, we should consider how much was paid for a typical poison, arsenic. During the 19th Century, in Yorkshire and Lancashire, one ounce of arsenic could be purchased for two-pence in old money (2d). A pound of sugar cost around four-pence (4d) and the average income per household was little more than thirteen-shillings (13s = 156d) a week.

In those days the infant mortality rate was appalling, with many, especially the new-born, dying due to suffocation, neglect, malnutrition, cruelty, violence and poisoning, including opiate overdose. The risks were high for legitimate children, but for those born out of wedlock the chance of survival was desperately low.

Rebecca Smith was found guilty of murdering her month-old son. She had produced a total of 11 children, but only the first-born was still alive. After her trial the authorities decided upon exhuming the bodies of the other nine children. Arsenic was found in some of them, and she confessed. It became clear that she was motivated out of concern for them, believing that they were better off dead than having to suffer and possibly die of starvation. She was probably correct in her belief, considering the poverty and appalling living conditions that the children were faced with. Although many thought she should not hang, the death penalty had to be imposed due to the number of murders. On 21st August 1849 Rebecca Smith was hanged.

It is curious that, not wanting any more children, a woman would go on to have another 10 or so. Perhaps she didn't know what it was that was causing the pregnancies. Or maybe her husband should have gone to chapel and found a clue in the reading from Genesis. *So when Onan went into his brother's wife, he wasted his seed on the ground.*

Rebecca Smith's case may well have been motivated by a sense of compassion, but in most child poisonings the motivations were personal gain. We see this in the case of Elizabeth Berry, a Liverpool workhouse nurse, who murdered to collect the insurance. The daughter of William Dawson Holgate died under suspicious circumstances, and he collected the £20 insurance. And Martin Slack poisoned his baby daughter with acid to avoid having to pay maintenance.

1.14 SELF-POISONING

Many of us accept that we are poisoning ourselves through our lifestyle choices. For example, we may choose to smoke, take drugs, drink too much alcohol or eat to the point of obesity. The poisoning isn't obvious at first, but in the long term we experience the pains of our indulgent ways. Here we have a case of 'chronic' poisoning. The opposite of this is 'acute' poisoning. For those who have lives that seem worthless, desperate or painful, poisoning is often the choice for ending it all. Thus, we see that those poisons that generate organ failure are chosen. A consideration is that the poison is fast acting.

The following are commonly used by those attempting suicide: pesticides; drug overdose; and carbon monoxide from a car exhaust. Not all suicides involve individuals, for there are mass suicides such as the Jonestown deaths. In this, people took a lethal concoction of diazepam and cyanides. Some would argue that the Jonestown case was not strictly suicide. Was it the decision of each individual to kill him or herself, or was there brainwashing and coercion?

It often happens that, in attempted suicides, the person does in fact survive the poisoning to be left with brain damage and all manner of other conditions that make life even worse than it was before their suicide attempt. Clearly most people do not opt for killing themselves and so live on until nature takes its course. But even here, as we move from life to death, poisoning plays a part. Poisons are involved in changing the body's chemistry in the process of dying. When the heart stops beating, the blood comes to a standstill. There is no more oxygen and glucose for cell respiration. The toxic products from the last traces of respiration build up around the cell and stop it functioning. Our cells are poisoned. Death has arrived. The machine grinds to a halt.

At that point, the body is but biological bits waiting to be broken down, first by its own enzymes then by earth or by fire. A corpse is dismantled and recycled. Its atoms and molecules are put back into the appropriate natural cycle – the nitrogen cycle, the carbon cycle and the water cycle. There they will wait until called upon to build a new living systems. Life goes on.

1.15 PUSSIES, PUPPIES AND POISON

Pets, especially when they are just starting to find their feet and are excited about exploring the world around them, frequently find objects to chew to see what it tastes like. If it is agreeable, they swallow it. Some pets will even swallow items that cannot be chewed down to bite-size chunks, and will swallow the whole item. The Labrador puppy is well known for this. Older pets may also resort to the same activity when they are bored, as the experience of chewing relieves boredom or loneliness.

Exploring new territory, new materials and new experiences is always risky, and many pets come to grief in the process. With regard to pet poisoning, there are dangerous chemicals in both the natural world and the man-made world. Indoors, many pets are at risk from chewing, and they may attempt to eat medicines that are still in their foil packs, plastic bottles and tubes. Some of these may be prescription drugs, dangerous to the pet even when only small amounts are ingested. A recent problem is when a pet bites into a refill cartridge for an electronic cigarette. This contains a solution of the poisonous alkaloid nicotine that can be as much as 0.15%.

Confectionary is appealing to dogs, especially when there is the smell of chocolate. Eating chocolate may not be a problem for us but for a dog it can cause serious illness, or even be fatal, as it contains the chemical theobromine, which is a potent alkaloid closely related to caffeine. Despite its name, the theobromine does not contain bromine. Houseplants and garden plants, such as the crocus and the rhododendron, are often a cause of death to the inquisitive pet.

Outdoors there are even greater risks. A cat licking that pond of water on the driveway may be poisoning itself. Despite the water tasting sweet, it contains antifreeze that often drips from the cooling system of a car's engine. The antifreeze is based on ethylene glycol, which after ingestion is converted into oxalic acid that destroys normal kidney function. Cats are more at risk than other pets as their kidneys have a low tolerance for some chemicals. This feline sensitivity explains why cats can be made ill if they are given certain convenience foods that contain preservatives such as compounds based on benzoic acid.

Slug pellets can be picked up in the garden and eaten, or ingested when the cat eats a dead mouse that has been poisoned by the pellets. The poison in the pellets is metaldehyde, which, in addition to molluscs, kills birds, cats and dogs. Another pesticide, bromodiolone, is a modern rat poison that domestic animals may access with fatal results.

Commercial pet foods have also posed a risk. In 2007 wheat gluten adulterated with melamine caused the deaths of many cats and dogs in America. The cause of death was kidney failure resulting from the melamine forming an insoluble substance that blocked the capillaries.

1.16 FARMERS' FEARS

Cattle are at danger from lead poisoning if they gain access to ground where waste has been carelessly dumped. Examples of the type of rubbish that pose a problem are lead batteries, sump oil and old paint tins. Cattle will also lick the flaking lead paint from surfaces, as they enjoy the sweet taste that the lead produces. The signs of intoxication start with a change in behaviour, with the affected animal becoming depressed and over-reacting to touch and sound. They become blind, stagger and crash into obstacles.

Cattle eating brassica plants may suffer nitrate poisoning, which can be fatal. Chemical fertilizers such as ammonium nitrate have also been known to cause nitrate poisoning. And synthetic pesticides such as the organophosphates may be fatal if ingested. Poisonous chemicals ingested in small amounts may seem not to affect the animal, but may bio-accumulate and end up in the food chain for human consumption.

The yew tree, often found growing in the churchyard, and the subject of much folklore, has lethal concentrations of taxines, toxic alkaloids, in its needles and berries. Cattle have died after eating clippings carelessly left on the ground. And the acorn from oak contains tannins that are toxic and cause kidney damage.

Young cattle often suffer when grazing in the pasture where there is ragwort. Although cattle and horses are at risk of ragwort, the grazing deer are unaffected. The poison in ragwort is an alkaloid that causes cirrhosis of the liver. Dropwater (water hemlock) is also a worry, for its exposed roots, called 'dead man's finger', contain the poison cicutoxin. Cattle also eat acorns and conkers, which contain a chemical that is broken down in the animal's body to tannic acid and gallic acid, both of which cause fatal kidney damage. Bracken can also kill cattle when they eat it, as it contains cyanogenic glycosides that break down in the animal's digestive system to cyanide ions.

Birds sometimes eat lead shot from clay pigeon shooting when they mistake it for grit. And lead shot may be found in the carcass of a game bird that is to be prepared for eating. Non-fatal poisoning of squirrels from chilli put in bird food may give the squirrel a bad experience and make it ill. The capsaicin in the chilli is what generates the burning sensation that many of us enjoy, but which other animals suffer from. But for the birds enjoying their meal of chilli, the capsaicin has no adverse effect.

Blister beetles, which are sometimes found in hay, can poison cattle. A part of the problem here is that during the haymaking process, the hay is compressed, and this crushes any beetles hidden in the hay. When the blister beetle is being crushed it releases large amounts of its defensive poison cantharidin.

1.17 GENDER FAVOUR

Combing through the texts on poisoning, we form the impression that it was the popular choice for a woman intent on removing an unwanted husband from her life. Even today there is a tendency to regard poisoning as a feminine approach to murder. In contrast we think of the masculine options to be stabbing, bludgeoning or shooting.

Severing an artery by stabbing or slashing results in blood being splattered all around, as the wound squirts blood on each beat of the heart – until it beats no more. And bludgeoning the head with a blunt object until all life is extinguished requires strength and stamina and also splashes blood around. Shooting is noisy, messy and leaves evidence in the form of a bullet lodged in the victim, or in the wall behind him, along with a very noticeable entry wound.

Certainly, for a woman to get rid of her husband, she is well placed in terms of opportunities if poison is to be her choice of weapon. It will be easy for her to slip something nasty into his glass of Rioja or prepare the latest Jamie Oliver concoction with an additional ingredient – a pinch or two of exotic spice. Despite our idea that poisoning is a woman's weapon, modern crime figures tend not to bear this out, as there seems to be little bias either way. But might there be other factors involved? Have cases of death by stealth slipped by undetected? Maybe women were more successful at poisoning because they were better placed to administer poison. Or are women more accomplished in activities that require an element of stealth?

Poisoning is a most subtle means of murder; it is not demanding in terms of physical effort or strength; it doesn't make a bloody mess; it makes no noise. Furthermore, the symptoms often mimic gastric complaints. It may be performed quickly, as in acute poisoning, or slowly, as in chronic poisoning. In the latter instance, the poisoning process is slow, and may pass as being due to natural causes. Of course, some poisons are a little less subtle and can arouse suspicion. Using strychnine, as we saw earlier, goes somewhat further than gastro-enteritis in terms of symptoms. It is hardly likely to be explained away as having eaten a dodgy kebab. But it has been known for strychnine poisoning to be dismissed as the result of drinking too much alcohol.

This section would be incomplete without mention of Britain's biggest serial killing. Over a period of years, Dr Harold Shipman murdered selected patients by injecting them with overdoses of diamorphine. On 31st January 2000 Shipman was convicted and sentenced to life imprisonment.

CHAPTER 2

Chemical Chaos

The trio of substances, strychnine, arsenic and cyanide, favoured by crime writers, are not poisons outright. Taken in small quantities, we find that, far from killing someone, they may have a beneficial effect. For example, strychnine was available as a medicine and a general tonic known as Easton's Syrup and arsenic was the active ingredient in the popular tonic sold as Fowler's Solution to cure sexual impotence. And we have all eaten, and possibly enjoyed, cyanide in the form of marzipan. In these examples the amounts usually ingested are sub-lethal.

2.1 DOSED TO DEATH

If we were to eat a 1 kg of marzipan in one go, it would provide a lethal dose of cyanide. But who in their right mind would attempt this? If we tried it, the chances are that we would vomit and thereby remove it from our system, as the body's response relates to the size of the dose.

Sprinkling a small amount of salt (sodium chloride) onto our food enhances its flavour and, so long as the amount stays small, there is no problem. However, applying larger amounts leads to a high concentration of sodium in the bloodstream. This causes an increase in blood pressure that eventually damages the heart, leading to premature death. In this example, salt is behaving as a

Poisons and Poisonings: Death by Stealth
By Tony Hargreaves

Published by the Royal Society of Chemistry, www.rsc.org

chronic poison, and it can take years before the damage is done. Ingestion of salt can also kill in a relatively short time.

The extent to which the body is affected depends not only on the quantity but also on the frequency of the dosage and how quickly it becomes active in the body, the 'bio-availability'. For a scientific study of poisoning, we must measure those variables. Although poisoning has been written about since ancient times, there was no systematic study until the 16th Century, when the physician and alchemist Paracelsus stated that all substances are poisons if taken in amounts above a certain level. Today this is quantified in terms of the lethal dose rating, LD_{50}.

For medicines, it is the dosage that distinguishes between poison and remedy. The amount for a remedial dose can sometimes be worryingly close to that of the lethal dosage – something most of us are aware of when taking paracetamol for a headache. All substances will kill if the body is made to absorb a sufficiently high level. Further progress in the understanding of poisons came in the 19th Century when Matthieu Orfila, physician to Louis XVIII, described the correlation between the chemistry of a poison and the human body's response to it.

2.2 ACUTE OR CHRONIC

In discussing poisons, we need to distinguish between rapid poisoning (acute) from a large single dose, and long-term poisoning (chronic), which may involve small amounts over years. Some chemicals, for instance arsenic, can play either role depending on how they are administered. In general, a poison is a substance that, when taken in a small amount, is absorbed by the body and changes the vital chemistry in such a way that death or injury results.

Choosing to reduce our dietary salt intake is one approach to healthier living. An alternative is to use low-sodium salt in place of ordinary table salt. The low-sodium salt that is currently available is simply a mixture of potassium chloride and sodium chloride. Potassium chloride is innocuous when ingested, even in quite large amounts.

If, however, potassium chloride was to be dissolved in water and injected into the bloodstream by means of a hypodermic syringe, cardiac arrest would result. The effect is rapid and certain, which is why it has been used for execution by lethal injection. In considering whether a chemical is poisonous or not, we must not only think in terms of quantities but also take account of the means through which it gains entry to the human body.

For a chosen substance to work as a poison, it must gain access to the body's organs by an appropriate route. Some substances may be versatile enough to be effective through a variety of routes. Many of the poisonings we look at in these chapters are planned killings.

This now looks like preparing to give a patient his medicine. In a way it is. It is the administration of a certain chemical in a manner that will cause changes in the body resulting in a particular outcome. The common ways in which a poison gets into the body are swallowing, injecting, inhaling and absorption through the skin. Most poisonings, especially suicides, involve swallowing the substance or inhaling the carbon monoxide from the exhaust gas of a car.

The lethal dose (LD) of many chemicals has been determined in animal experiments, and so the figures obtained are only rough approximations for comparing poisons in the human body. Some examples of lethal doses are shown in the Glossary.

2.3 SOLUBLE CYANIDE

Substances that are water-soluble solids can be easily dissolved in the victim's drink or in their food during its preparation. Cyanide salts have often been administered by such means. However, not all cyanide compounds are water soluble, and so they do not form aqueous solutions. This does not mean that they fail to dissolve and thereby do not become bio-active. They may well dissolve in the body's fatty tissue, and create havoc from there.

If a solution of sodium cyanide is swallowed, it instantly enters the bloodstream and goes on to poison the various cells by disrupting their respiration. Swallowing sodium cyanide as coarse crystals would result in a delayed reaction, as the crystals would take time to dissolve in the body and form a solution. This is due to the large crystals having a smaller surface area than the same weight of fine crystals. Surface area is the controlling factor because it is at the surface where the dissolving occurs.

When we ingest cyanide in the form of sodium ferrocyanide it has no effect. Most of us do this on a daily basis, as the chemical is put into table salt to prevent it going lumpy in damp conditions and keep it as a free-flowing powder. We are not poisoned, as the compound is not broken down. No free cyanide is formed, and the chemical passes through the digestive system and is then excreted.

The solubility of the actual substance is of importance when considering its poisoning potential. Potassium cyanide has a higher solubility than that of its sodium counterpart. Thus, potassium cyanide dissolves more quickly than sodium cyanide. Marzipan contains cyanogenic glycosides, which break down in the digestive system to form free cyanide ions that enter the bloodstream. However, this amounts to only minute levels of cyanide. When eaten in normal amounts, the cyanide never reaches the lethal concentration.

Arsenic, generally as white arsenic, has a long history of being administered orally. It is noteworthy that dispersing the arsenic in a clear or colourless drink produces a cloudy suspension that would hardly pass unnoticed. However, if left, a small and sufficient amount of arsenic will slowly dissolve and the liquid will return to being clear again. And if the drink is acidic the white arsenic will dissolve more effectively.

Attempts to murder by administration through swallowing have often been defeated by the victim's natural defence mechanism. When the stomach receives something nasty, vomiting will occur until the offending chemical has been completely expelled. The arsenical dish prepared with so much care for an unwanted husband ends up all over the carpet. Hubby lives on and decides to opt for takeaways for future meals. Back to the drawing board for the budding poisoner.

Assuming no vomiting occurs, the poisonous substance now resides in the stomach. Depending upon the type of poison in question, the stomach acid may chemically change the poison, making it more or less active, or it may totally destroy it, turning it into harmless chemicals. This is why some therapeutic drugs cannot be taken in tablet form, but have to be administered by injection. In addition to the stomach acid, there are also digestive enzymes that can destroy certain poisons. For example, snake venom is a protein that is broken down by the digestive enzymes into harmless amino acids.

If the poisonous chemical resists the reactive conditions in the stomach, it may then pass to the small intestine. This organ has a huge surface area, approximately 300 m^2 in an adult, which absorbs nutrient molecules and transfers them to the bloodstream along with the poison. However, different types of substances are absorbed at different rates. Substances that are fat-soluble are often absorbed more quickly than those that are water-soluble.

Chemical names can cause confusion, so we need to be careful. For example, mercury chloride comes in two forms. There is mercuric chloride, which also goes under the name of 'sublimate of mercury' and is now called 'mercury(II) chloride'. Then there is mercurous chloride, which was called 'calomel' but is now called 'mercury(I) chloride'.

The difference between these names may seem trivial, and they are both composed of the same elements, but the poisonous properties are seriously different. The mercuric chloride is extremely poisonous because it is highly soluble and so releases large concentrations of mercury ions into the bloodstream. Mercurous chloride has low solubility and so releases only minute amounts of mercury ions. The naming of chemicals is further discussed in the Glossary.

Also, the time taken for the poison to present itself in the bloodstream is dependent upon factors that can change a situation from one that is fatal to one that is survivable, and results in nothing more than a bit of tummy ache. Was the poison taken on a full stomach or an empty stomach? What is the body mass of the victim and the general state of health? A glass of Rioja laced with poison may well be fatal if gulped down before the meal, but may pass unnoticed if slowly sipped during all three courses.

2.4 INJECT OR INHALE

We see that the venomous snake has evolved to inject its venom into a victim's flesh by means of fangs. Injection is a highly effective way of administering a poison, and if that injection is intravenous so much the better. Intravenous injection typically allows the whole dosage of poison to enter the bloodstream in the time it takes to discharge the liquid from the syringe – a matter of seconds. There is no delay as there would be if the poison were to be ingested and time was taken for it to be absorbed. Injection, especially if it is intravenous, produces a rapid reaction.

However, injection normally requires the co-operation of the person who is to receive the dose. This is acceptable if the substance to be injected is therapeutic and administered by medical personnel. Suspicions may be aroused, however, if a husband finds out that his wife has a secret stock of syringes and needles and has taken an interest in the veins in his arms.

For this reason, we find that injection of poison is often the domain of doctors and nurses who are at liberty to inject people at will. No questions asked. Here, the name Dr Harold Shipman springs to mind, for he committed a series of murders by injection of high doses of diamorphine. Killing a person by means of intravenous injection is also used for the lawful processes of euthanasia or execution.

Similar to intravenous injection is the intravenous drip that supplies medication constantly, often over a period of hours. Adding poison to the intravenous solution has been tried by some killers. Feeding tubes are sometimes required to take nutrients directly into the stomach and offer another means of poisoning administration, as we see in the case of Susan Hamilton who poisoned her daughter with salt by this means.

Inhalation is the route to absorption where the poison is gas, vapour, or dust. Once again, the absorption challenges of the digestive system are circumvented. Inhalation of dust may make it bio-available through the nasal membrane. It is this mechanism that enables drugs like cocaine to be taken, and is commonly referred to as 'snorting'. Although little practised these days, tobacco in a form called 'snuff' was also taken by this method. A poisonous vapour, once it has reached the alveoli in the lungs, becomes active as it enters the bloodstream. For those who smoke, the immediate effect of inhalation is noticeable; that first inhalation of a lit cigarette produces a response within a second or two.

Suicide by inhalation is common these days because the poisonous gas required, carbon monoxide, is readily available from the car exhaust. Accidental poisoning occurs due to inhalation of organic solvent vapours and aerosol propellants, especially among young people who enjoy the narcotic effect of inhalation.

2.5 SKIN DEEP

Some poisons gain access to the body by absorption through the skin and mucous membranes of the eyes, nasal passage and the uro-genital parts. We tend to think of our skin as being a waterproof barrier, and from this comes the idea that it will also prevent substances other than water entering the body. This is generally true, but there are some chemicals that easily pass through the skin and become distributed in the body. Many of the chemical warfare gases work on this principle.

Methyl mercury is treacherously poisonous by skin absorption because it rapidly dissolves in fats and is therefore absorbed by fatty tissue in the skin. Benzene, a hydrocarbon solvent, was at one time used for cleaning and degreasing but has now been withdrawn due to its capacity for absorption through the skin and into the body where it can cause leukaemia. Back in the days of witches, flying ointment was applied to the skin, enabling certain chemicals to be absorbed into the body and become active, as we shall see later.

Absorption of poison through the mucous membranes is also common. Examples are the absorption of snake venom from the spitting cobra that directs the poison into its victim's eyes, causing blindness. We have all noticed the effect upon the eyes of chopping onions, an effect which is the basis of tear gas. The genitals may also absorb poison, but the fact that they are generally covered lessens the chance of them being affected. Whilst in the uro-genital department, it is worth noting here that an enema is a means of introducing a poison, as we shall see in the Marilyn Monroe case.

2.6 DISTRIBUTION AND EXCRETION

Once inside the body, the poison is delivered to different parts by the bloodstream. As the poison moves around, various organs become exposed to it. Damage may be caused to a number of them but in general there is one particular organ where the action of the poison is quicker or more intense. It is as though a particular type of poison aims for a specific organ, the target organ. It may be the bloodstream itself that is damaged by the poison. For example, the red blood cells that carry the oxygen can have their haemoglobin poisoned by carbon monoxide. Or the aqueous fluid that the blood cells are transported in may also be damaged, such as when formic acid, from methanol poisoning, enters the blood.

Arsenic, in addition to attacking the intestinal tract, can also end up bound to the protein in hair and nails. Other heavy metals behave likewise. This has proved useful in post-mortem examination of bodies to establish the cause of death, as we shall see later when considering the role of arsenic in Napoleon's death.

Many poisons are detoxified by the liver, but dealing with large doses may damage the organ. For example, the liver will happily deal with a normal dose of paracetamol, but supply it with an overdose and its normal chemistry will be seriously changed with fatal consequences. In general, the liver will break down organic molecules such as paracetamol, but it cannot deactivate poisons such as arsenic, antimony and lead because these are elements. These toxic elements cannot be broken down by any chemical means. Only nuclear reactions in atomic bombs or atomic reactors can split the atom that makes up the particular element.

Excretion is typically through urination or defecation. However, a poison may be removed by other means, such as vomiting, perspiring, exhaling, or being locked away in hair, nails and bones so that it is biologically unavailable. The amount of time the poison is present in the body before it is broken down or excreted varies from poison to poison.

The usual way of measuring the amount of time it takes for a poison to be broken down is to use the concept of 'half-life'. Consider a poison being broken down in the liver. At the onset of the poisoning, the poison will be at its highest concentration, and the rate of breakdown will be high. Sometime later, when a proportion of the poison has broken down, the rate will be less. When most of the poison has been processed, the rate of breakdown will be very slow.

We see that the rate becomes slower and slower, and the actual point at which it stops cannot be measured accurately. Half-life is a measure of the time it takes, under specified conditions, for a poison to be reduced to half its original concentration, and it can be measured with reasonable accuracy. Use of half-life is a standard means for assessing poisons. Examples of half-life values are shown in the Glossary.

2.7 POISONS' REACTIONS

The human body is a complex system of millions of chemical reactions perfected by millions of years of evolution. These are not independent reactions; many of them run in series; some run in parallel; some do both. An enormously complex matrix of interconnections exists. For any chemical reaction to work, it must be constantly supplied with reactants and have its products removed. When either supply or removal stops, the chemical reaction reaches equilibrium, the balance between reactants and products.

To all intents and purposes, the reaction has ground to a halt. Each reaction is a part of a system in which co-ordination is the key to proper working. The failure or disruption of one reaction cannot, therefore, occur as an isolated event. There is always a knock-on effect. Thus, exposing the body to a poison typically disrupts one of the reactions, and as that reaction fails, subsequent reactions follow suit. A simple analogy is the derailment of an express train. Derail one carriage, and it's not long before the whole train comes off the track, with disastrous consequences.

The rail-crash inquiry will call for a sifting through of the carnage and wreckage in the hope of learning where the initial disruption occurred. Was it accidental or intentional? It's not unlike the way forensic scientists go about examining a poisoning victim.

In our efforts to understand how poisons work, it is essential to know what the first step was that initiated the damaging sequence. Below, we consider some poison chemistry in detail to examine the primary poisoning step, *i.e.*, the reactive stage once the poison has arrived at its target. Generally, the attack begins with a poison interacting with a cell or the fluid around the cell. Here are a few of the more common poisoning mechanisms, seen in terms of their chemistry.

The reactions that follow are simplifications of a few selected mechanisms. They have been chosen to show the vital connection between chemistry and the human body. Poisoning mechanisms may just involve a single type of reaction, or could also consist of a complicated sequence of processes.

2.8 CHELATES

Oxygen transport in the blood relies upon haemoglobin picking up oxygen molecules in the lungs and delivering them to the cells. Haemoglobin is a large molecule that contains an iron atom bonded in a special way to form what is known as a 'chelate'. The iron atom in the chelate attaches itself to oxygen from the lungs and carries it, in the bloodstream, to the body's cells, where it releases it. The oxygen-bound molecule is called 'oxyhaemoglobin'. A critical factor is that the chemical bond between the oxygen and the iron atom in oxyhaemoglobin is weak, and so the oxygen is easily released.

However, if carbon monoxide enters the lungs and comes into contact with the haemoglobin, it forces its way into the chelate, producing a chemical bond that is 200 times stronger than that for oxygen. An irreversible situation results, in which a new molecule, now known as 'carboxyhaemoglobin', is formed. The carboxyhaemoglobin cannot carry oxygen to the cells, nor can it release the carbon monoxide. The cells are starved of oxygen, and, in turn, respiration shuts down and death results.

2.9 SEQUESTERING

The special chemical bonds that form the chelates mentioned above can be used to trap poisonous metal ions that have entered the bloodstream. For example, Lewisite, also known as The Dew of Death, is a volatile arsenic compound that was used in chemical warfare. The compound worked by the arsenic combining with proteins in the body's cells and destroying their proper functioning, resulting in damage to exposed parts such as eyes, skin and lungs.

Dimercaprol was developed as the antidote. It works by forming a chelate with the arsenic from the Lewisite, and thereby sequesters the arsenic. By such means, the arsenic is prevented from reacting with the proteins, and it is effectively neutralised and then excreted. Dimercaprol is also used today as an antidote for poisonings caused by antimony, lead and mercury, since it can sequester all these elements in a similar way.

2.10 pH BALANCE

Acids chemically react with bases during neutralisation reactions. In the human body, the essential chemical reactions take place in water. An important property of an aqueous solution is the pH value, as this measures precisely the changes in the acidity of the solution. Consider the blood, for example. It has evolved to function correctly when its pH is 7.4 because this is the pH resulting from a certain concentration of bicarbonate ion arising from the transport of dissolved carbon dioxide to the lungs for it to be exhaled.

We would expect an acid (pH less than 7) entering the blood to cause a lowering of the blood's pH, and *vice versa* for a base entering the blood. However, this does not happen, at least when small amounts of acids or bases find their way into the blood, because the bicarbonate acts as a buffer. Any small change in pH is compensated for by this buffering action, and the pH soon bounces back to normal. Of course, if large amounts of acid enter the bloodstream, then the buffer capacity is overwhelmed, and life-threatening changes occur in the pH of the blood.

It is possible for formic acid to be produced from the breakdown of certain poisons in the liver. When the acid leaves the liver and enters the bloodstream, the blood becomes acidic. It no longer functions correctly and death is likely. It is worth noting that the formic acid's poisonous action is not confined to the blood, but also attacks the optic nerve. There may be insufficient formic acid to kill, but enough to cause blindness.

Although it is most unlikely that anyone would attempt poisoning – murder or suicide using concentrated acids such as sulphuric acid, it is possible that someone might accidentally swallow some of the liquid. The experience would be horrendous, and the vomiting mechanisms would be primed in an instant. Although the acid may never make it as far as the bloodstream, it would have done a lot of corrosive damage to the mouth and oesophagus. Strong bases, such as caustic soda solution, have a corrosive effect similar in many respects to that of sulphuric acid.

2.11 CRYSTALLISED KIDNEYS

Two dissolved chemicals may react together to form an insoluble solid. If the insoluble substance is formed rapidly, it comes out of solution as a fine powder, known as a 'precipitate'. Where the insoluble substance forms slowly, it appears as crystals that often cluster as they come out of solution.

For example, when oxalic acid is ingested, it remains in solution and soon enters the bloodstream where it encounters dissolved calcium ions. A reaction occurs to form calcium oxalate, which is insoluble and comes out of solution as a precipitate. It is this precipitate reaction that occurs in the kidneys and can damage the filtration process. If the reaction occurs slowly, large crystals form, and these also cause damage or even complete destruction of the filtration system. Oxalic acid and the problem of forming insoluble calcium oxalate can also occur in the liver and other organs, with serious consequences.

Arsenic and other poisonous metals react with enzymes in the body fluids, and these reactions also form precipitates. In this process, the enzyme is effectively removed from its working environment, and a vital reaction ceases with fatal consequences. Arsenic in the digestive system forms a yellow precipitate of arsenic sulphide that coats the inner surfaces of the intestines. In the days when arsenic poisoning was commonplace, this was often noted during post-mortem examination of the victim's internal organs.

Another process that blocks the kidneys is that of crystallisation, and this occurs with oxalic acid playing a slightly different role to the one outlined above. If a high concentration of oxalic acid is produced, it may remain dissolved until the fluid in which it is dissolved also has other soluble chemicals added to it. The result is that the least soluble chemical, for instance the oxalic acid, crystallises out and forms blockages.

The formation of solid clumps of material often occurs quite naturally in the human body. The working of the body relies upon the flow of blood from one place to another through capillaries, veins and arteries. Fatal blockages can occur as a particular organ shuts down due to failure of supply of essential substances. The body may well survive the failure of a particular organ for a short time. However, if the organ in question is the heart then a shut-down does tend to throw a spanner in the works and wreck the entire machine.

A well-known example is a blood clot that blocks the heart's main artery causing the heart to stop beating. There is an interesting twist here in which a deadly poisonous chemical, warfarin, supplied as a non-lethal dose, prevents a heart attack. This is because the chemical acts as an anticoagulant, and artificially thins the blood, enabling it to flow effectively.

2.12 ELECTRONS AND IONS

Acetylcholine is a chemical that enables nerve impulses to pass from nerve to muscle, causing the muscle to contract. After passage of the impulse, the acetylcholine is destroyed by enzyme action. This facilitates the switching on and off of the nerve cells. If, however, the chemical is not destroyed, the impulses continue and the muscles contract uncontrollably.

Strychnine, and related poisonous alkaloids, block the enzyme, and control of the electrical signals fails. The muscles to which the signals are sent begin contracting uncontrollably, resulting in the convulsions that we observe in a victim of strychnine poisoning. Warfare gases such as nerve agents also work by attacking the enzyme that breaks down acetylcholine.

Another effect that has an electrical basis is due to cyanide. Cyanide, usually in the form of the sodium or potassium salt, quickly and easily passes through the cell membrane to create havoc. Once inside the cell, the cyanide heads for the part of the cell in which respiration occurs to produce energy. This is a complex reaction involving electron-transfer processes. The presence of cyanide disrupts these, and the body's energy production ceases, resulting in convulsions leading to death.

The aqueous fluid in which the blood cells are suspended contains various dissolved ions known as the 'electrolyte content'. A correct balance in the concentrations of these ions is essential for the correct functioning of the body's cells. For example, a bicarbonate ion is needed to transport carbon dioxide from the cells to the lungs; sodium and potassium ions maintain the critical concentration across the cell membrane; calcium ions are carried to the bones; ammonium ions are taken to the kidneys to be filtered out and excreted.

Changing the electrolyte balance upsets the working of the body. This can occur when water acts as the poison. If huge amounts of water are ingested, this is exactly what happens. The water content of the blood increases, resulting in dilution of electrolytes, and the balance is upset. Death results due to insufficient electrolyte concentration. It is surprising that water, a chemical that accounts for 70% of the human body mass, can act as a poison, but it is simply an example of Paracelsus' statement.

For the cell of a particular organ to function properly, requires chemicals to pass in and out through the cell membrane. For instance, in the respiration reactions that produce energy, molecules of glucose (the fuel) and oxygen pass into the cell through the cell membrane. Waste products, carbon dioxide and water, pass out of the cell *via* the cell membrane. The cell membrane is therefore important in regulating the movement of chemicals in and out of the cell. A vital part of the functioning of the cell membrane relies upon potassium ions being transported through it.

These potassium ions are supplied by the bloodstream. If, instead of potassium ions arriving at the cell membrane, thallium ions are delivered, then the chemistry goes awry. The thallium ion has the same size and same charge as a potassium ion. This enables it to mimic potassium and sneak into the membrane. Thallium, like many other heavy metals, forms a strong chemical bond and attaches itself to proteins, preventing their proper functioning. The cell membrane therefore fails to work, and the organ is irreversibly damaged.

2.13 SOLVENT AND SOLUTE

Solvents such as chlorinated hydrocarbons, ketones and esters are attracted to fats, and so accumulate in the body's fatty tissue. Inhalation of solvent vapours, as occurs when young people sniff glue or the contents of aerosol cans, results in the fatty tissue absorbing some of the solvent. Much of the solvent will come out of the fats when the source of exposure is removed. However, there are certain organs, for example the kidneys, in which the solvent not only is absorbed into the fat but also dissolves and removes some of the fat and destroys the normal functioning of that organ.

Nerve cells are especially prone to solvent attack. These cells carry vital electrical signals. They have a fatty sheath that acts as an electrical insulator wrapped around the conducting part of the cell. When the solvent molecules reach the sheath, they are absorbed. The result is that the effectiveness of the insulator is reduced and the nerve cells fail, causing short circuits between adjacent cells. A particular example of this is when the solvent reaches the fatty tissue of the heart muscle, and prevents correct functioning resulting in a life-threatening reaction.

2.14 BLOCK THE BLOOD

One natural component of the blood is vitamin K, and this is needed to coagulate the blood where damage to a blood vessel results in loss of blood. Without a coagulating mechanism, the person would bleed to death due to even the slightest of injuries whether external or internal.

Vitamin K acts as the coagulant, but in the process the vitamin K molecule becomes oxidised. Once the process is complete, the oxidised vitamin needs to be chemically reduced to make the vitamin K available for further use. Warfarin inhibits the enzyme that performs the reduction process, with the result that the internal bleeding is not blocked but continues. The victim thus dies due to internal haemorrhaging.

2.15 TOXIC OXIDE

Some chemicals entering the human body are not, in themselves, poisonous, but are changed by reactions in the body to chemicals that are poisonous. This commonly happens in the liver. One example is when the poisonous alcohol, methanol is ingested. When the methanol reaches the liver an enzyme oxidises it to formic acid, which then enters the blood, overwhelms the blood's buffering capacity, and causes the blood to become acidic. When a certain level of methanol is ingested, normal functioning ceases and death results.

Ingestion of ethanol from alcoholic beverages also results in oxidation of the alcohol by an enzyme in the liver. In this instance the acid produced is acetic acid, but this is a normal part of the body's chemistry, and the acid easily undergoes further oxidation to carbon dioxide and water.

Swallowing antifreeze containing glycol results in the liquid being carried by the bloodstream to the liver. Once there, it is oxidised by an enzyme to produce oxalic acid. The oxalic acid enters the bloodstream and, as mentioned earlier, causes precipitation of calcium oxalate, which is insoluble, and blocks the kidneys with fatal consequences.

2.16 PROTEIN AND RICIN

Ricin, the deadly poison from the castor oil plant, is a large molecule made up of two protein chains. In the victim, it binds to cell membranes where it starts to split into the two separate chains. One chain migrates into the cell and attacks enzymes in the part of the cell responsible for protein synthesis. With no more protein being produced, the cells, particularly those of the liver and kidneys, fail. Over a period of hours, multiple organ failure occurs, and internal haemorrhage results. The death of the vital organs is accompanied by blood in the vomit and the diarrhoea.

The lethal dose for ricin is below 10 micrograms per kilogram of body weight ($\mu g\ kg^{-1}$). As such, a 70 kg person is killed with as little as 0.0007 g of ricin. In the tiny pellet that was used to murder Georgi Markov, the amount of ricin present was about 0.0004 g. This is, of course, the amount of ricin remaining after the poisoning, *i.e.*, after much of it had been released into Markov's body.

2.17 ANTIDOTES

Nowadays all hospitals are equipped to deal with poisonings, and for the bigger hospitals, poisoning cases arrive on a daily basis. Most of these are accidental poisonings where children have swallowed tablets, household chemicals, berries and seeds. Many of the adult poisonings are suicide attempts. With old people, poisoning is usually due to them becoming confused over their medication. Occupational poisoning of people in industry and poisoning due to bad food or drink is less common.

In general, deliberate poisoning as part of an attempted murder is rarely encountered in hospital emergencies.

Prompt treatment of the patient in hospital usually results in recovery. The general approach is to keep the patient alive until the poison is either excreted or broken down by the liver. If the poison is highly toxic, it is often removed by emptying the stomach by gastric lavage. If the patient is at home and seriously ill from the poison, it may be decided to induce vomiting as a last resort. Here however, there is a risk of patients choking on their vomit or inhaling their vomit and thereby carrying the poison into the lungs.

Giving the patient activated charcoal is now a common treatment, when it is thought the poison is still present in the stomach. In the stomach, the charcoal comes into contact with the poison. The poison is strongly adsorbed onto the surface of the charcoal and rendered harmless. Eventually, the charcoal passes through the digestive tract and is excreted with the faeces.

Activated charcoal has a surprisingly high capacity for adsorption, especially where the poison is an organic compound. For example, 1 g of charcoal will absorb about 0.25 g of ethanol. However, for salts, acids and bases, charcoal is less effective and in some cases useless. If the poison has been absorbed into the bloodstream, then blood filtration by means of dialysis may be used, especially where the kidneys have stopped working. Other means of blood filtering are available depending upon the poison involved.

Some poisons have antidotes that are effective if administered promptly. However, the medical staff will need to know exactly what has been swallowed and have an indication of the quantity. Dimercaprol is used to treat poisoning by the metals antimony, arsenic, mercury and thallium. It is a sequestering agent and locks on to the offending metal, preventing it reacting and causing damage. The metal is tightly bound to the Dimercaprol and is eventually excreted. Another treatment for heavy metals poisoning is to administer calcium-EDTA by intravenous injection. As with Dimercaprol, the poison is sequestered and later excreted.

Injections of atropine are used for organophosphate pesticides and some nerve agents. For poisoning due to morphine and related narcotics, injections of naloxone are used as the antidote.

In cases of carbon monoxide poisoning, usually by intentional inhalation of car exhaust fumes, oxygen is given. For paracetamol poisoning, the antidote acetylcysteine is effective if administered in the early stages.

People who work with cyanide have antidote kits available in case of accidental ingestion of powders or solutions, or inhalation of dusts or hydrocyanic (prussic) acid gas. If the latter is inhaled, the poisonous action is so fast that there is unlikely to be enough time to use the antidote kit. These kits contain amyl nitrite, which is inhaled to stimulate the heart, and intravenous injection materials for administering solutions of sodium nitrite and sodium thiosulphate. Cyanide poisoning can be effectively neutralised, but only if treated promptly.

CHAPTER 3

Animal and Vegetable

The natural environment produces an enormous range of poisons from the living species that inhabit it – plants, animals, fungi and bacteria. A large proportion of poisonings involve plant toxins. However, many of these poisonings are mild and, in general, they do us no more harm than simply causing short-term pain or minor illness. We have all experienced, for example, the discomfort from stinging insects or stinging nettles, and soon recovered with no lasting effect apart from a heightened awareness for the offending species.

With plants, animals and fungi we can see the enemy, but with bacteria there is nothing to be seen. There's little, if anything, that can be done to avoid a sore throat due to bacteria invading the area and causing us to suffer the discomfort. The release of bacterial toxins causes the ill effects until our immune system, or our prescription antibiotics, destroy the offending micro-organisms. Of course, it is wise to avoid situations that we believe are favourable for bacteria to thrive in.

With fungi, we are mindful of the need for great caution to distinguish between edible mushrooms and deadly toadstools (Figure 3.1). In fact, such is the risk that, without expert knowledge of fungi, it is generally wise to regard them all as poisonous and not go picking our own, despite the temptation of fresh food free for the taking. The relatively mild pain from grazing past

Poisons and Poisonings: Death by Stealth
By Tony Hargreaves

Published by the Royal Society of Chemistry, www.rsc.org

Figure 3.1 Fly agaric is the fairy tale toadstool. It contains the toxic chemical muscin, which in small amounts makes the victim delirious. Larger amounts cause hallucinations, nausea, diarrhoea, convulsions and death.

some stinging nettles is a discomfort that lasts for only a short time, whereas the wasp sting is seriously painful and lasts longer. And in hotter countries a poisonous snake is given a wide berth for fear of it injecting its venom into us. We might ask ourselves why plants and animals produce poisons. Clearly, defence is one role, and this becomes evident when we consider the actions of venomous animals, for, in general, they do not attack unless they feel threatened. However, there are some animals that produce venom in order to kill their prey.

Prickly thorns and stinging hairs of the plant kingdom are surely a means of deterring a hungry creature from making a meal of the plant. There are plants and animals that contain poisons that stay within them, but upon being eaten, or even chewed, the poison takes its effect. The result may be fatal or it may be no worse than a nasty taste in the mouth. Some plants are toxic because poisons from the soil, such as lead, cadmium, fluoride and selenium, are absorbed along with the normal plant nutrients. Fungi can be particularly efficient at absorbing heavy metals from the soil. We also need to be wary of eating vegetables grown on soil that contains poisonous chemicals, either natural or from, say, industrial waste of the past. What might seem like a nice healthy cabbage might well have absorbed something sinister into its foliage.

Over millions of years, humans have evolved alongside poisons. In fact, a part of the evolutionary process has equipped us to cope with many poisons. However, it has not prepared us for poisoning that is planned and executed by our fellow beings. Although there are many poisons available from living systems, they are often not readily accessible. Animal venoms are difficult and dangerous to collect. Imagine the methods that would need to be used to collect venom from snakes, scorpions, spiders and jellyfish.

Furthermore, the venoms would normally have to be injected or introduced onto the mucous membranes of the victim to be effective. Swallowing them would, in most cases, be useless. Most venoms, being a mixture of proteins and enzymes, would be destroyed in the stomach by the action of the hydrochloric acid and the stomach's own proteolytic enzymes that are needed for digestion.

For those animals that have the poison within their skin, consideration may be given to butchering followed by adding the fleshy bits to the victim's food or wine. But, however much effort was expended in the preparation, it is likely the victim would become suspicious on having a strange experience of trying to chew the rubbery bits of brightly coloured skin.

Insects may also be considered, but their size makes it a major effort to collect enough of them. Going down further in size, there are the bacteria. These days we know enough to be able to cultivate bacteria but, on completion, some kind of vehicle is required for the victim to absorb the bacteria. Despite these difficulties, bio-poisoning through the growing of bacterial cultures has been attempted, but only to a minor extent.

When all is considered, it is the plant toxins that take first place in nature's range of poisons when it comes to poisoning. Furthermore, many plant toxins are available, not as poisons but as chemicals with legitimate uses. It is no surprise to find poisons of plant origin playing a major role in murder.

A wide range of creatures use venoms for defence or for attacking their prey. Examples of species that are venomous include snakes, ants, bees, scorpions, spiders, jellyfish, stingrays and stonefish. Some of these may not be venomous enough to kill a human, but they certainly create pain in varying degrees. An attempt to quantify the intensity of the pain produced the Schmidt pain index. The index goes from zero to four. At zero there is no pain; a wasp or bee sting is rated at level two. Anyone suffering the excruciating pain that level four represents must wonder if life is worth living.

3.1 VIPERS AND VENOM

In terms of the fear factor, the snake (Figure 3.2) comes top of the list, and not without reason, for it is estimated that deaths from snake bites amount to around 100 000 a year.

A venomous snake with its lightning-fast reactions is well equipped to deliver its poison. With its hollow fangs for piercing skin and injecting the poison, it is efficient. Many of the victims it claims are taken unawares, and so it is no wonder that the snake terrifies many. Snake venom is saliva that has been modified to contain toxins needed to attack and immobilise their prey. The venom also contains enzymes that will work on

Figure 3.2 The venomous snake, with lightning-fast reaction, delivers the whole dose of poison in a fraction of a second by means of hollow fangs, which pierce the victim's skin and inject the venom. Many of its victims are taken unawares – death by stealth. The venom is saliva containing toxins and enzymes. The toxins immobilise the victim and the enzymes begin the process of digestion of the victim.
© Shutterstock.

the dead or immobilised prey and begin the process of digestion by breaking down the victim's proteins, fats and carbohydrates.

Vipers seem to be especially well equipped for delivering their venom, which, once injected into the prey, attacks the blood circulatory system, causing coagulation and clotting in the arteries. With regard to venoms being used to carry out murder, there are cases in which a venomous animal, usually a snake, is placed in a prime spot by the culprit. However, a poisoner planning the direct use of animal venom is faced with some serious difficulties. Venoms are produced in specialised glands, and delivered by means of spines, teeth or other devices for piercing the skin of its aggressor or its prey.

Collecting the venom is bound to be hazardous, and administering it to the victim poses problems because, for it to be effective, it generally needs to enter through pierced skin or, for some venoms, *via* a mucus membrane such as the eye. Furthermore, the amount of venom available for collection may be minute. The difficulty in collecting animal poison is in sharp contrast to the ease of extracting poisonous chemicals from plants. Administering venoms by means of swallowing is likely to have little effect, because venoms are usually proteins that simply end up being digested by the enzymes in the stomach with no ill effect.

Attempting suicide by means of venom reminds us of Cleopatra's efforts with an asp. The snake was likely to be the Egyptian cobra, and whether Cleopatra's death was suicide or murder is debateable. An analysis of the story considers the forensic questions and was recently reported in the *Sunday Times*.

3.2 PUFFER POISON

Some animals contain chemicals within their flesh that are not strictly venoms but are toxic to those who eat the flesh. It seems strange, therefore, that we choose to eat such creatures. The globe fish, also known as the 'puffer fish', is a well-known example. It is a delicacy in Japan, where a type of soup known as 'fugu' is made from the fish.

Properly prepared by expert hands, dishes containing globe fish are safe, but, when incorrectly made, the dish can contain the neurotoxin tetrodotoxin. Most of the toxin is present in the ovaries and liver of the fish. In small doses it causes a tingling sensation in the mouth, nausea, headache and loss of coordination. In larger amounts the toxin causes convulsions and death.

3.3 POISON FROGS

There are many species of frogs (Figure 3.3) in the rainforests of Columbia. The golden poison frog, which is possibly the most poisonous of animals, has enough poison to kill 10 grown men. It is 1.5–6 cm in length, and produces the poisonous alkaloid batrachotoxin from glands beneath its skin. Contact between human skin and that of the frog is sufficient to cause poisoning. The neurotoxin is absorbed, leading to paralysis and death, as vital functions of the nervous system become blocked.

Hunters in the South American forests hunt birds and monkeys using darts laced with the poison. The hunters pierce the frogs with sticks, resulting in an injury that causes the frog to release the poison as a froth. Arrows are then dipped in the froth

Figure 3.3 In the tropical forests of Central and South America there are many examples of frogs. Their brilliant and beautiful colours are meant to warn predators. Glands on the back of the frog secrete highly toxic substances that can poison someone simply by skin contact. The frogs do not make their own poison, but accumulate it from their diet of ants and termites. Hunters in the forests use the poisonous secretion to put on the tips of blow-darts.
© Shutterstock.

and allowed to dry. Other poison frogs have different alkaloids as poisons. For example, the alkaloid epibatidine is produced by an Ecuadoran frog commonly known as the Anthony's Poison Arrow Frog. This poison causes brain damage, respiratory paralysis, seizure and death.

In general, poisonous frogs are brightly coloured (red, green, yellow, blue) to deter predators. The brighter the colour, the more poisonous the frog. If the predator ignores the colour and attempts to eat the frog, the poison is released into the predator's mouth and the frog has a chance of escape. It may well have suffered an amount of chewing and be in poor shape, but at least it stands a chance of surviving.

Predators, assuming they are not killed by the poison, quickly learn to avoid these frogs. Poison dart frogs do not actually make the poison, but often accumulate it from their diet of beetles, ants and the toxic insects to be found in the rainforests. Dart frogs and puffer fish have the most lethal toxins of all species, but they are not venomous creatures, and transfer their poison only when in contact with a predator.

3.4 ERECTION SPIDER

There are tens of thousands of species of spider, but very few of them cause deaths. However, there are some notable ones that strike fear into many people who live in hot countries. The Brazilian wandering spider is venomous and highly aggressive. Its venom is 20 times more powerful than that of the black widow spider. The venom contains a neurotoxin that blocks the calcium channels of the nerve cells. Serotonin is released, causing intense pain and breathing problems due to loss of muscle control, leading to paralysis and death by asphyxiation.

A non-fatal bite, although painful, can result in an erection that may last for several hours. There is currently interest in this aspect of the venom by researchers studying erectile dysfunction. This spider is also known as the 'banana spider', and there have been many cases of boxes of bananas accidentally carrying the spiders around the World. Recently, a woman in Swansea opened a box of bananas bought from a local supermarket, and was shocked to see the fruit had a large cocoon of spider eggs.

Shortly after opening the sealed box, the cocoon began to open to release the baby spiders. The bananas, from Costa Rica, caused a certain amount of fear. The main concern was that some spiders may have escaped and were hiding in their new-found home. In this case, no-one was poisoned, and the supermarket agreed to a refund so long as the bananas were returned with their barcode from the packaging. Despite the potency of this Brazilian spider, and its ability to wander Worldwide, there are few fatalities, as an antivenom is available.

3.5 BOX JELLYFISH

The box jellyfish is transparent and almost invisible in the sea. It has tentacles that produce nemacysts. This is a complex mechanism for the delivery of the toxin. When the tentacle is touched, a trigger device causes a hinged mechanism to release a barb with enough force to penetrate the skin of the victim. The barb is connected to the tentacle by means of a long thread, which then pumps the toxin, a hypotoxin, from a capsule into the skin of the victim. Prior to triggering, the thread is coiled. The toxin causes excruciating pain and may be fatal.

The box jellyfish is featured in the Sherlock Holmes story *The Adventure of the Lion's Mane*. The story relates to Holmes, whilst relaxing on a Sussex beach, coming across a man in agony who mumbled the words "lion's mane" before he died. The man wore only overcoat and trousers. He was seen to have red welts on his back and over one shoulder that looked as though he had been thrashed with a cat o' nine tails.

The Portuguese man o' war is similar to the box jellyfish in that it has venomous tentacles that can deliver a painful sting, which can be fatal. It uses these to kill fish, which, once stunned, are pulled into the digestive system, where enzymes break down the proteins, carbohydrates and fats to feed the man o' war.

3.6 BLISTER BEETLE

There are many species of the blister beetle. They have a length of about 1–2.5 cm, and most are brightly coloured as a warning to predators. When the live beetle is crushed, it exudes cantharidin from its joints to defend itself and to protect its eggs, which are coated with the chemical. In the dried-out body of the insect there is a high concentration of this chemical, which is a powerful vesicant – hence the name 'blister beetle'. A powder made by grinding the dried bodies has a long history of use for treating ailments.

Today, cantharidin preparations are used in medicine to a small extent; for example, in removal of warts, skin blemishes and tattoos, as cantharidin is powerful enough to burn off human skin. For this application, the chemical is collected from a particular type of beetle that is generally known as 'Spanish fly', which can yield high concentrations of the chemical. When ingested, cantharidin irritates the urinary tract and may cause swelling. In a woman it causes a tingling effect in the urethra when she urinates. In a man it results in a condition known as 'priapism', in which the penis takes on an erection that can last for several hours and become extremely painful. It is these effects that led to the erroneous idea that cantharidin is an aphrodisiac. More serious symptoms from ingestion of cantharidin are internal haemorrhage after intense abdominal pain as the kidneys are destroyed and the victim urinates blood.

Knowing something of the so-called 'aphrodisiac' properties of cantharidin but nothing of its treacherously poisonous properties, Arthur Kendrick Ford decided to try it out. He obtained some from his workplace with the intention of giving it to three of the women at work, in the hope of them becoming sexually aroused. Presumably he thought that they would turn to him for sexual gratification. After placing a large dose of powdered cantharidin into some sweets, he offered them to the women who enjoyed what they believed to be ordinary sweets.

The cantharidin set to work, but the dose was so large that any effect upon the genitals would have been overpowered by the torture of the blister agent. Within a few hours the women were suffering horrendous pain as the cantharidin slowly attacked their insides. Two of the women died, and a post mortem revealed the corrosive action of the chemical.

When interviewed by the police, Ford confessed to having given the women sweets that he had spiked. It was clear that he had no intention to kill them, and as such he was convicted of manslaughter rather than murder in 1954. It is worth noting that there are sex products advertised with a Spanish fly theme. These do not contain cantharidin but other herbal ingredients that are harmless – and quite possibly useless for their stated purpose. Clearly, a substance as dangerous as cantharidin would never be sold to the public.

In 1772 the French aristocrat Marquis de Sade is rumoured to have given aniseed pastilles containing Spanish fly to two prostitutes at an orgy. It seems he was motivated by a desire to increase his sexual pleasure. The women survived the experience. He was sentenced to death for attempted murder, but later won an appeal and was released.

3.7 HOT BEETLE-SPRAY

Another beetle that uses chemical defence when it is threatened is the bombardier beetle. This is not poisonous like the blister beetle, but has an interesting weapon. It ejects a spray of hot liquid from its abdomen, which is sufficient to deter an aggressor.

The force to drive out the liquid comes from the decomposition of hydrogen peroxide, releasing oxygen, which causes pressure to build up. At the same time, enzymes catalyse a reaction between hydroquinone and more of the hydrogen peroxide. This reaction produces 1,4-benzoquinone and a large amount of heat, which raises the temperature to almost boiling point.

A valve in the abdomen then opens, and the foul-smelling hot liquid is released with such explosive force that it produces a distinct popping sound. The bombardier beetle has a sufficient supply of chemicals to enable it to release about 20 blasts of liquid, enough to kill a small predator (Table 3.1).

3.8 MALICIOUS MICROBES

There are many micro-organisms that, once they've set up home inside us, release toxins that can make us ill or kill us. Bacteria pose the biggest risk. They are often present in the food we eat, the water we drink, the air we breathe, and we may also be exposed to them through people and animals we have physical contact with. When present in the air, they are carried from place to place in dusts and spores. Minute particles of water that have become dispersed in the air as aerosols may also transport

Table 3.1 Other animals that are known to be venomous or poisonous.

Type of animal	Animal/location/venom
Snake	(a) Black mamba. Africa. Most feared of all snakes. Fast-acting venom releases 400 mg each bite. Highly aggressive. These are serious poisoners. They have evolved for death by stealth when taking their prey or killing a potential threat. Fast and efficient snakes that are feared worldwide.
	(b) Indian cobra. Most rapidly acting venom of all snakes. Hundreds of thousands of bites each year, with average 15 000 annual fatalities. Fatal within 45 minutes but can be treated with prompt action.
	(c) King cobra. Southeast Asia. Produces largest volume of venom at 1000 mg. Has mild temperament making it less aggressive than many other snakes.
	(d) Spitting cobra. Africa and Asia. Spits venom rather than injecting. Aims at eyes. Affected eye suffers corneal swelling due to neurotoxin. Can cause permanent blindness.
	(e) Taipan. Australia. Most toxic venom of all land snakes. Can yield 100 mg of venom, which is toxic enough to kill 100 men. Venom contains batrachotoxin, which attacks the victim's nervous system.
Frogs and toads	(a) Corroboree frog. Produces own poison rather than absorbing it from food source. Pseudo-phrynamine. A psychedelic, used in religious rites.
	(b) Colorado River toad. Releases bufotoxin from glands on back.
	(c) Asiatic toad. Toad venom. Used in oriental medicine.
	(d) Common toad. Releases bufotoxin from glands on back.
Spiders, scorpions	(a) Black widow spider.
	(b) Deathstalk scorpion. Pain and respiratory failure.
	(c) Fat-tailed scorpion.
Sea creatures	(a) Blue-ringed octopus. Most venomous animal in the sea. Skin contact with this octopus results in blindness, respiratory failure and death.
	(b) Stonefish. Camouflaged among stones and rocks. Carries toxin on its back. Skin contact causes paralysis.
Birds, geese	(a) Birds. Capped ifrit and pithui. Batrachotoxin, absorbed from insect diet, is concentrated in skin and feathers. Contact produces numbness and stinging.
	(b) Spur-winged goose. Cantharidin in its tissue from diet of blister beetles.
Butterflies	Monarch. Poison absorbed in diet from its caterpillar days when feeding on milkweed. Toxins are the steroids, cardanolides.

bacteria. Some years ago, the public health campaigns would inform us that "coughs and sneezes spread diseases".

3.9 THE BLACK DEATH

One of the worst cases of death by infection with bacteria was due to bubonic plague (Figure 3.4), commonly known as 'The Black Death'. It killed an estimated 100 000 000 people as it spread throughout Europe between 1346 and 1353. This made it the greatest poisoning ever to be caused by toxins released by invading bacteria.

It was death by stealth on a pandemic scale. These toxins caused horrific symptoms, starting with a rose rash that formed a ring. Next came fever, headache, aching limbs and delirium. Swelling of the lymph glands to form buboes in the neck, armpits and groin followed. These buboes then began to ooze pus, and parts of the body began to turn black as they died and started to decompose. The overall effect of the bacterium is to poison the cells' ability to connect with the immune system, thereby rendering the body defenceless.

Figure 3.4 The headgear used to protect against the evil air during the time of the plague that arrived in London. As the disease spread throughout Europe in the mid-1300s, it claimed an estimated 100 000 000 lives. The arrival of the disease in Britain was believed to be in the form of a virus carried by a flea, which lived on rats that infested the cargo holds of ships.
© Shutterstock.

Recent research suggests that the bacterium responsible was *Yersinia pestis*, and was carried from the Orient by fleas living on the black rats that had found their way onto merchant ships. The fleas then spread to humans. The playground singing game comes from the time of the Black Death. A small group of children form a ring to dance in a circle around a single child known as the 'rosie'. On reaching the last line of the song, the children would all fall to the ground. The slowest child to fall then became the next rosie. There are slight variations from one country to another; the British version is as follows.

Ring-a-ring o' roses,
A pocket full of posies,
A tishoo! A tishoo!
We all fall down.

3.10 IN-HOUSE BUGS

The Black Death involved bacteria from far-off lands reaching us to create havoc. However, closer to home we have bacteria ready to invade us and release their deadly poisonous proteins and enzymes. Our cosy warm homes with their fitted carpets and central heating are a haven for malicious microbes. Unless carpets are washed regularly, they harbour bacteria that are happy to poison us if they get the chance. *Escherichia coli* and *Salmonella* in carpets is a recently recognised source of infection.

Carpets also become populated by dust mites, resulting in a build-up of their faeces, which contain dangerous bacteria. Washing the carpet minimizes the problem, but how many people do this? When it gets filthy, they pull it up and lay a new one. But in the process of removing the dirty carpet, the room becomes filled with dust, which is inhaled. A big risk with carpets is that babies crawl on them and maybe pick up interesting things to chew, and thereby ingest the bugs.

All of us have all suffered poisoning from these microorganisms, and in most cases we survive the assault when our immune system is activated and comes to our defence, or when we take a course of antibiotics. When we suffer infection by a virus, there may be no cure, and so it is a matter of waiting and suffering until the virus moves on to another host.

Live bacteria produce and release toxins, which are mixtures of proteins and enzymes, into the body, causing poisoning. Dead bacteria that have been killed by the body's immune system release toxins as they disintegrate. Toxic chemicals from these bacteria make us ill by disrupting the normal cell functioning or by actually killing the cells. The ways in which the damage is done involve: (a) breaking open the cell membrane, allowing the contents to leak out; (b) blocking protein synthesis within the cell; and (c) preventing the proper working of the cells at the nerve junctions.

To appreciate the extent to which pathogenic micro-organisms cause human suffering, we find examples on a regular basis in the news. Some of the more important cases are considered below. As with all other poisonings, the victims are unaware that something that is life-threatening has gained access to their body through food, water, air or contact with an infected person or animal. Nothing appears to be wrong until the symptoms kick in, cause havoc, and take us to death's door. Once again we see death by stealth.

3.11 FOOD POISONING

Many cases of food poisoning are caused by pathogenic bacteria, and most of us have suffered from this. When we eat food contaminated with bacteria or toxins from bacteria, we become ill. It is not simply that the bacteria that have gained access to your body are having a poisonous action. The real problem is that the bacteria, being living organisms, reproduce at a rapid rate. The few bacteria that infected us an hour ago had a negligible toxic effect, but now there are many millions. Now the food poisoning symptoms show, as the digestive system fails to work properly.

Typical symptoms include nausea, vomiting, abdominal pain and diarrhoea. In most instances the poisoning is uncomfortable and soon passes, but, in some cases it proves to be fatal. Bacteria frequently responsible for food poisoning are: *Salmonella, Listeria*; *Escherichia coli*; and *Campylobacter*. And viruses, such as the norovirus, may also invade us and start to build up a huge population that, in a few hours, makes its presence known. Contamination of food sources occurs by several means; the main ones are: poor food hygiene procedures; vegetables fertilised with animal manure; and shellfish from water containing sewage.

3.12 CANNED CORNED BEEF

The Aberdeen typhoid outbreak in 1964 resulted in over 400 being admitted to hospital with symptoms of food poisoning. The source of the poisoning was traced to corned beef produced in South America. The meat was from a single large tin, and was then used for cutting into slices in the food shop.

The slicing machine then contaminated with typhoid bacteria came into contact with other products through poor hygiene procedures. Investigation revealed that, during the canning process, the meat had come into contact with cooling water that had not been chlorinated to kill off bacteria. The source of the water was the Uruguay River, which was polluted with human faeces from untreated sewage. Thanks to a prompt response from the health authority in Aberdeen, there were no fatalities, and all those poisoned returned to normal health after a stay in the hospital. Typhoid fever is caused by the *Salmonella* bacteria, and most people recover naturally. In the most serious cases the disease can progress to erosion of the intestinal wall, leading to haemorrhage.

The case of Mary Mallon, who became known as Typhoid Mary, is well known in the history of this disease. Mary was immune from typhoid, but was found to be a carrier of the disease after investigation by the health authorities during an epidemic in New York in 1904. With the authorities in pursuit, she disappeared and continued with her work as a cook, spreading the disease further afield.

Eventually they caught up with her and committed her to an isolation centre, but subsequently released her when she agreed no longer to work as a cook or in a similar position involving the handling of food for the public. Once free, she broke the agreement by again working as a cook, and another outbreak of typhoid started. The authorities returned her to the isolation centre where she remained for the rest of her life. Typhoid Mary caused many cases of typhoid, and three deaths were directly attributed to her.

3.13 OFFENSIVE OYSTERS

Even with a dedication to top-quality food, bugs can still cause poisoning by stealth. A restaurant in Britain was in the news in 2009 when 240 cases of nausea and vomiting occurred. Health officials discovered that oysters contaminated with the norovirus were the likely cause.

Oysters, being filter feeders, are prone to accumulating bacteria. In this case, the oysters from a source in Essex were contaminated with bacteria from human sewage. The supplier of the oysters had also sold them to other restaurants, which resulted in a total of 529 people being made ill.

Norovirus, commonly known in Britain as the 'winter vomiting bug', affects over 250 million people Worldwide, resulting in around 200 000 deaths a year. The symptoms of this highly contagious virus are nausea, vomiting, abdominal pain and watery diarrhoea. With good healthcare services, most people recover after a few days, but in less-developed countries many people die from the virus. The virus is killed by heat and chlorine-based disinfectants.

The problem with the oysters was that they were being eaten alive, just as they should be. There would have been no suspicious taste. If they had formed part of a cooked dish, there would have been no problem. If you eat living oysters from a reliable source, accept the very small risk of being poisoned. But not all sources are reliable, as in many parts of the World raw sewage is still discharged into the sea.

Microbiologists estimate that fewer than 20 virus particles are enough to cause infection from the aerosol produced when an infected person vomits. In this respect, projectile vomiting can be a serious problem, as those nearby may believe they were well clear from the vomit, but unknowingly some aerosol particles have already infected them. Even flushing a toilet in which there is vomit or diarrhoea can produce the aerosol particles that carry the infection.

3.14 COOKED HAM

A butcher's shop sold contaminated ham, resulting in 21 cases of poisoning from a strain of *Salmonella*. The ham came from a supplier whose other customers, in various parts of Britain, experienced a similar problem. *Salmonella* poisoning from food is usually caused by eating contaminated raw or under-cooked food that is already carrying the bacterium. Meat, eggs and dairy products are common sources of *Salmonella* (Table 3.2).

Table 3.2 Poisonous microbes.

Bacterium/ virus	Year/location/source of infection/number of deaths
Escherichia coli	(a) 2011. Germany. Fenugreek sprouts. Worst case of *E. coli.* Caused 53 deaths.
	(b) 1993. America. Undercooked hamburgers killed four.
	(c) 2006. America. Spinach. Three were poisoned.
	(d) 1996. Juice from rotten apples caused one death.
	(e) 2005. Britain. Meat. One fatality.
	(f) 1996. Britain. Cold cooked meat. Caused 21 fatalities.
Listeria	(a) 1985. America. Cheese. Worst *Listeria* case. 50 people died.
	(b) 2011. America. Melon. Killed 30.
	(c) 2008. Canada. Cold cooked meat. 22 died
	(d) 1998. America. Hotdogs. Poisoned 20 people.
	(e) 2014. Denmark. Sausage, bacon, liver, paté. Denmark's worst case. Killed 15 people.
	(f) 2002. America. Contaminated poultry. Caused eight deaths.
Salmonella	(a) 1985. America. Milk. Worst contaminated milk case. Caused nine fatalities.
	(b) 2008. America. Peanut butter carried infection and caused nine deaths.
	(c) 2015. America. Cucumbers from Mexico took the lives of four people.
	(d) 2016. Britain. Ground cumin and coriander.
	(e) 2006. Britain. Chocolate. A million chocolate bars were recalled from sale.
Botulinum	1963. America. Canned tuna contamination resulted in two deaths.
Campylobacter	2009. Britain. Chicken liver.
Hepatitis A	2003. America. Green onions.

3.15 FOOD FRAUDSTERS

In addition to bacteria and viruses poisoning us, we may also be poisoned by food that contains chemical adulterants, which are substances added during the manufacture. The motive is usually to boost profit, but sometimes adulterants are put in to compensate for an ingredient being below par.

Industrial food, manufactured by the ton using ingredients from many different suppliers in many parts of the World, is a growing public health worry. Often the ingredients have a less-than-perfect pedigree, and some are of dubious quality. We are also faced with the rising problem of food fraud, especially where high-cost ingredients are used. Some of the cheaper brands of convenience foods are nothing more than flavour-enhanced, colour boosted, industrial slop made from the poorest ingredients. Standing on the supermarket shelf in its glossy packaging it looks delicious and gives the impression that all the boxes have been ticked. But what lingers behind those respectable facades?

As some supermarkets found to their embarrassment, their beef burgers contained horsemeat. Their trusty suppliers, it seems, had not been as thorough as they should have been, resulting in the horsemeat scandal of 2013. It was a case of adulteration, with the motive being financial gain by unscrupulous people in the supply chain.

Those of us who ate burgers in 2013 were left wondering. Was it good horsemeat? Or, was it pet-food grade from an Eastern European knacker's yard, and condemned as unfit for human consumption? Did it contain veterinary antibiotics? Was it from a diseased animal? Did a quality inspector mark the right box: 'tick for pass' to accept; or 'tick for fail' to reject? In this horsemeat case the primary adulterant was, at face value, not a serious issue. After all, many people throughout the World eat horsemeat. But, the horsemeat in this scandal was put into the food by stealth. Who knows what else it brought with it?

3.16 FORENSICS FINDS FOUL FOOD

It makes financial sense for the food fraudster to adulterate the costlier foods. That's where the real profit is. For example, it has recently been found by forensic chemists that saffron, the most expensive spice in the World, has been adulterated with sandalwood dust, yellow dye and gelatine.

And some other scams recently uncovered are: corn oil sometimes has the cheaper rape seed oil added; papaya seeds are mixed in with peppercorns; olive oil is diluted with tea tree oil; pure orange juice may not be quite so pure as sometimes it is adulterated with grapefruit juice; and apple juice from certain sources has been found to contain synthetic malic acid. We might ponder over the description '100% pure orange juice from concentrate'. What was added to the concentrate to make it into the juice. Tap water with chlorine as disinfectant, maybe? The list goes on and grows each day.

Adulteration and related food scams are seldom fatal, but can be poisonous enough to be damage our health. But we must be on our guard as fatalities do occur; for example, when red lead was added to poor-quality paprika to boost the colour, when melamine was put into baby milk so that the quality analysis was fooled, and when white wine had ethylene glycol added to improve sweetness.

Food forensics is growing, and it must because the modern open markets for trading are a free ticket to fraudsters. The production of food is commercial, it is done for profit, and if the profit is low, or the company is in a bad financial way, then, as we say, "anything goes". If the worst comes to the worst, then it is common practice to 'take the risk, grab the cash and make off'. In a typical large town, restaurants come and go with alarming regularity.

In the past, people attempted to ensure their food was as fresh as possible. They had to, for they did not have the luxury of chilling, freezing, freeze-drying and canning. It is ironic that in modern times we eat food that may be years old.

Poisonous chemicals in food may be there by accident. There was concern some years ago about mercury in tuna fish. The problem here was that mercury had moved up the food chain and, in so doing, had become more concentrated in the living tissue. The tuna fish ate salmon, which ate krill, which had absorbed mercury from a river carrying mercury-polluted effluent to the ocean. The mercury in its progress up the food chain was bio-accumulated. Thus, dangerous levels were found in the tuna. But tuna was not the only risk, for shark and halibut were also affected, as they typically feed on tuna.

The krill had only a small concentration of mercury in their organs; the fish further up the food chain had a far greater concentration. It is safe for us to eat quite large amounts of krill, but we must eat only a small amount of tuna, shark or halibut.

3.17 FOUL WATER

A putrefying corpse releases a cocktail of bacteria and decay products. The human body is 64% water, 20% protein, 10% fat, 1% carbohydrate and 5% mineral matter such as calcium phosphate. After death, it takes little time for nature's processes to begin the decomposition so that the chemicals in the dead body are recycled.

We need to consider the events that occur in nature when an animal dies. It typically collapses onto the ground, and before long the scavengers are at work. They move in and cautiously, for there are other hungry beasts about; they set to work eating the soft tissue. Soon, all that remains of the corpse is a skeleton, teeth, claws and hair. The process of destroying the flesh is relatively short. The skeleton remains and is eventually, maybe hundreds of years later, mineralised and becomes plant nutrients.

In the absence of scavenging animals, the soft tissue remains to be broken down by putrefaction, then fermentation, which is a lengthy process. A human body is subject to the same processes of degradation. If left to nature, the body is quickly and cleanly disposed of, and its component chemicals are on their way to make up new living systems.

However, human bodies are not left on the ground, and are treated in a way that nature did not intend. With our inventiveness, we humans have devised a whole range of activities to deal with this death and decay. The most common is burial, but this action simply increases the amount of time taken for the body to decompose. In cooler climes, a body buried beneath about 3 m of soil has a slow rate of decay, taking around 10 years to reach the skeletal stage. As such, our burial grounds must accommodate huge numbers of deceased before the ground can be re-used for further inhumation. In times gone by, when a cemetery was full, the skeletons were dug up and taken to the charnel house.

Cemeteries are no more than landfill sites of rotting flesh. And like all landfill sites, there is the problem of contaminating the groundwater during the length of time it takes for all the soft tissue to completely decay. Of course the rate at which a corpse decays is influenced by the type of coffin that contains it. In Britain there are about 200 000 burials each year, and most of these use coffins made of chipboard, which, in the wet earth, disintegrates within a week or so. With the fabric of the coffin having broken up, the body is exposed to air, water and earth. The normal process of decay then begins.

Some coffins are made of medium density fibre board (MDF), and some use wood, a popular type being elm. The latter takes a long time to break down, and can protect the corpse from being attacked by carnivorous creatures and decomposers such as external bacteria and fungi. Air is also excluded so that, once the body's own bacteria have used up the limited amount of oxygen contained within the cask with its screwed-down lid, their respiration is prevented and they die. As such, a corpse in an elm coffin can take as long as 60 years for the flesh to decay to leave only the skeleton.

In the past, important people such as royalty were buried in lead coffins whereas the lower classes were put into wooden casks, and the poorest of all were simply wrapped in a shroud and buried. The lead coffin would typically have its seams and lid soldered to make the whole thing airtight. In a lead coffin, the decay process would go so far, until the last traces of oxygen had been used up by respiration in the bacteria. Thus, the corpse remains almost intact, that is, until the archaeologist comes along, digs up the coffin and breaks into it to rummage around the rotting remains.

Whilst the coffin remains airtight and buried, its contents are unable to pollute the ground. It is likely that the bacteria have long since been killed by the poisonous lead, but they may have left spores that could be a problem if they are released. Outside the coffin, the lead is likely to have been in contact with wet soil, and traces of the metal will have dissolved into the groundwater, which may be someone's water supply. The potential for lead poisoning exists. Lead coffins are no longer used. Wood and cardboard are the preferred materials, and these will, in general, decay to harmless chemicals, mainly carbon dioxide and water. Unfortunately, the corpse encased within has an unwholesome decay profile.

Assuming there is sufficient air for bacterial decomposition, then the proteins and fats of the flesh will be broken down into simple chemicals such as amino acids and fatty acids that soak away in the surrounding ground water. The particular chemical reactions involved and the rates at which they occur depend on the soil conditions. Fungi will also play a part, and use the organic matter for their own respiration. Peaty ground, which is acidic, will have a different effect to ground in a limestone area, which will be neutral or slightly alkaline. Whatever the soil chemistry, the final products from the decay of flesh are mainly gases such as carbon dioxide, hydrogen sulphide, sulphur dioxide and ammonia. These would soon seep out through the ground and be released into the atmosphere.

The remainder of the decay products will be soluble substances that percolate downwards into the groundwater. Groundwater coming from a cemetery is likely to carry some nasty bugs that can kill if ingested. The Pennine village of Heptonstall is notable in this context. A well that supplied drinking water was located near to a cemetery. This was not unusual practice, and it would have posed no problem if the ground had been well drained. But the bedrock was near to the surface, and this prevented the water from the burial ground from quickly draining downwards to take it away. The result was that some water flowed sideways and into the well. In the 1840s Heptonstall's poison well led to 51 cases of typhoid fever. The bacterium causing typhoid is a strain of *Salmonella.*

In some countries such as America, there is the worry of arsenic poisoning the soil around cemeteries. In the past, arsenic was used in embalming fluids to slow down the rate of decay, so as to keep the body in good condition for the funeral. The arsenic was functioning as a bactericide and fungicide. After burial, arsenic leached into the groundwater as the body decayed. Modern embalming fluids do not use arsenic preparations, but instead rely upon formaldehyde, methanol and phenol, which are less of a problem in terms of poisoning the ground.

3.18 DISSOLVING THE DEAD

On a global scale, the figures show that each year about 55 million people die, and around 130 million births take place. As such, the disposal of the dead is a huge and growing problem. The issue of full cemeteries is becoming serious, as are the concerns about cremation. Burning the body takes an enormous amount of gas to reach the cremation temperature, which means more carbon dioxide being released. There may also be release of mercury from dental amalgams being vaporised. Volatile chemicals from wood preservatives in the coffin and from embalming fluid are also released.

Burial at sea has few pollution problems, and the flesh is devoured by the fish. Sky burial, as practiced in the Buddhist charnel grounds of the Himalayas, is similar, in that the body's soft tissue is eaten by scavengers and carrion eaters. In this instance, it is the vulture that gets a free meal. This method of body disposal is not without its problems. There have been occasions when birds of prey, having bitten off more than they can chew, attempt to fly off with their piece of flesh but lose their grip on it. Bits of human flesh falling from the sky onto the local villages causes concern, alarm and disgust.

There is less of a problem in those charnel grounds that cut the body into bite-size chunks rather than leaving it as a whole piece. Sky burial ensures the flesh, as protein and fat, is recycled within a matter of hours. It is the nearest we can get to a natural means of disposing of the dead as it is truly organic – the green approach. Pieces of bone, teeth, hair and nails remain to be slowly broken down by the bugs in the soil, the ultraviolet light from the sun, and the oxygen of the air.

Once again, human inventiveness comes into play. The latest idea is to dissolve the body using a potassium hydroxide solution at 140 °C. This breaks down the proteins into their component amino acids, and turns the fats into potassium salts of fatty acids. The latter is the basis of manufacture of soft soap from animal fat, a process known as 'saponification'. Advocates of the process, which is called 'resomation', are saying that it is environmentally beneficial in that it does not emit huge amounts of carbon dioxide during the burning of the methane to produce the high temperature for a period of hours. At first sight, all is well, but look more deeply and some nagging doubts occur.

How much carbon dioxide is emitted when the potassium hydroxide is manufactured? This bit of industrial chemistry uses electricity to make the chemical by electrolysis. As most electricity is generated from burning fossil fuel, this raises the issue of carbon dioxide emissions. A complex equation begins to take shape. We might look at this simply by thinking of carbon dioxide emissions of the actual process. But, to be totally honest, we need to consider the whole analysis and formulate a life-cycle assessment. Then, resomation loses some points.

Resomation may well destroy fat and protein, but leaves insoluble material from the calcium in the bones. In this respect it is little different to cremation. But, what happens to the brown liquid that was the soft tissue of the organs? Although there are suggestions that it may be used to put on the ground, this would require further processing to neutralise residual alkali that would poison the soil. The main method, and likely the cheapest method, of disposing of the foul fluid from the flesh, is to discharge it into the sewer. The thought of this does not sit happily in the mind. The organic remains from the loved one end up in the sludge tank at the sewage works with the mass of festering faeces. Once there, the amino acids and fatty acids are broken down by bacteria, and the carbon released to the atmosphere as carbon dioxide. Just as in cremation, the organic residue from the corpse ends up in the atmosphere.

By comparison, cremation and burial are respectful. In both cases, much of the organic material, the once-living parts, are released as gas to the air or as solutions to the soil. The body's spirit rising to the heavens, or its aqueous remains feeding the soil, is surely better than going down the sewer.

But we have to face the facts, and find new means of body disposal for a world that produces millions more dead bodies every year. The problem of dying is growing.

3.19 BOTULINUM FOR BEAUTY

Here we consider poisoning on the microscale. A bacterium, invisible to the naked eye, can produce a tiny amount of chemical that disrupts the normal working of the tissue that the bacteria have invaded. Bacteria reproduce so rapidly that, in the right environment, they can soon produce an enormous population, with each individual releasing its toxin into the host.

Intentional poisoning requires either the introduction of harmful bacteria into the victim, or the introduction of the toxin itself. To produce the neat toxin requires growing the bacteria in a culture medium and then extracting the toxin from that medium. These laboratory techniques require special equipment and trained personnel. Introducing the living bacteria or their spores to the intended victim is the other approach.

The toxin botulinum is a protein, and one of the most poisonous known. It is found in processed food such as canned meat, which has not been properly heat-treated. Its name derives from the Latin and means 'sausage poison'. Normally it lives in the soil, producing spores that are quite resistant to heat. If the spores are on the food to be canned, they may, once sealed in the can, release the bacteria that, in turn, release the toxin. Although the spores have good resistance to heat, the bacteria themselves are unstable to heat and are killed by cooking. Unfortunately, many of the canned products that the bacteria have thrived in are those that are eaten cold, straight from the can, thereby carrying the bacteria to the consumer.

We can appreciate the toxicity of botulinum toxin by considering the data: 1 g of the pure toxin dispersed into the air as an aerosol could kill a million people. When absorbed into the body, it affects the central nervous system. The symptoms of botulinum poisoning are: dizziness; blurred vision; vomiting; and restricted dry throat. Death occurs by paralysis of the cardiac and respiratory muscles. If caught in the early stages, botulinum antidote is effective, but the nervous system may already be damaged.

Botulinum has been considered as a bio-weapon since World War II. In the 1990s it was used by the Japanese cult Aum Shinrkyo in an attempted bio-terrorist attack. They tried to discharge botulinum toxin into the air, but failed due to faulty equipment that was supposed to release the toxin as an aerosol. The group had prepared the toxin from soil samples collected in Japan. Iraq in 1990 deployed specially designed missiles with a 600 km range; 13 of them were filled with botulinum toxin, 10 with aflatoxin, and 2 with anthrax. Botulinum has its legitimate medical uses, and is sold commercially in a highly dilute form for treatment of muscle over-action, such as squint, and to provide relief from migraine and muscle pain. One of the common brand names is Botox. Perhaps the better known application of Botox is in cosmetic surgery.

3.20 POISONS FROM PLANTS

The greatest number of plant poisons comes from a group of chemicals known as the 'alkaloids'. Alkaloids are chemically related to ammonia, which makes them behave as bases. Their function in nature seems to be defensive. Alkaloids are waste products from protein metabolism in the plant. It seems plants have evolved to use these waste products to stop the hungry herbivore driving them to extinction. When alkaloids from one plant accumulate in the animal eating it, a poisonous reaction is set up that makes that animal move to another plant. In small amounts, alkaloids stimulate and support the immune system of the eating animal by acting like a homeopathic medicine.

If animals and humans eat the same plant for a long time, they can suffer from alkaloid poisoning. However, this would be difficult to achieve in circumstances other than starvation rations. For example, if you had a diet that was exclusively spinach, it is certain that the accumulation of specific alkaloids would make you have no appetite for this plant. Alkaloid poisoning does occur when people try extreme diets. The first symptom is tingling finger tips, but the poisoning is reversible with no long-term effects.

In all, there are many thousands of different alkaloids, with some plants containing a range of them. For instance, the opium poppy contains 30 different alkaloids. Reacting as bases, alkaloids are easy to isolate from the plant material, which enables their purification. Both extraction and purification involve treating the plant material with acid, whereupon the alkaloid is neutralised to form a salt that can be crystallised out of the solution.

Nearly all the alkaloids produce some kind of physiological response once inside the human body. As such, they have played an important role in the history of medicine from early times. In fact, some still play an important role and are used alongside the modern medicines of today.

Some examples of plant alkaloids, and related compounds, used in medicine are: codeine, an analgesic; quinine, an anti-malarial; ephedrine, a blood-vessel constrictor; cocaine, a powerful local anaesthetic; and tubercurarine (curare), a muscle relaxant used in surgery. A few alkaloids are found in non-plant sources. For example, the arrow poison from the poison dart frog, and ergot from certain fungi.

3.21 STRYCHNINE

This is an alkaloid formed in various parts of the tree *Strychnos nux vomica*, which grows in India. The fruit of the tree contains almost 2% strychnine, and is well known for poisoning birds and rodents. Strychnine was readily available and in common use as rat poison. It was used in small amounts in medicines up until the early 1970s, and was regarded as a stimulant capable of increasing awareness. It was sold as Easton's syrup, which comprised ferrous phosphate, quinine and strychnine. Also available in sugar-coated pill form, strychnine was used as a nervine and general tonic.

These pills caused the deaths of children, even as late as the early 1960s, as they mistook them for sweets. Where the lethal dose was reached, the rapid poisoning effect would begin and bring death within about 15 minutes through convulsions. There is no known antidote, but for those surviving for three hours after exposure, a full recovery is likely.

Earlier we saw its use by Dr Thomas Neill Cream, a sadist and moral degenerate, who murdered prostitutes. Dr William Palmer also used strychnine in addition to other poisons, as we shall see later. Other infamous users of strychnine include Eva Rablen, who was convicted of the murder of her husband. Shortly after she had taken him a cup of coffee, he was found screaming and in convulsions upon the floor. Rablen died before medical attention could reach him, but in his last breaths it could be just made out that he was saying the coffee was bitter.

The post-mortem examination, including analysis of his organs, showed nothing unusual, and it was concluded that he had died of natural causes. Rablen's father was suspicious, and persuaded the police to make further enquiries. During a search, they found a bottle, labelled 'Strychnine', behind a broken plank.

The label also bore the name of the supplying pharmacist, who the police visited and learnt that the poison had been purchased only three days before the death. The stomach contents were re-examined, and this time strychnine was found.

Failure to find strychnine during the initial analysis was put down to the fact that an inexperienced assistant had performed the tests. Analysis for strychnine was then performed on Rablen's clothing, on the cup that had contained the coffee, and on the dress of a woman that Eva had spilled some of the coffee upon while she was carrying it to Rablen. Eva Rablen pleaded guilty at the trial in June 1929, and was sentenced to life imprisonment.

3.22 ACONITE

Aconite is found in the dried root of the plant wolfsbane, which has been grown in Britain since early Medieval times. It is an extremely poisonous plant, but butterflies happily feed on it without ill effect, and low-dosage preparations from this plant have a therapeutic value. The active compounds aconine and aconitine are alkaloids, and in small doses function as analgesics.

Aconite has a history of use in medicine going back to the early civilisations. In ancient Greece it was known as the Queen of Poisons. Nicander of Colophon described it as the "quickest acting of all poisons", even through skin contact, and especially through the mucous membranes. He noted that death occurs the same day if the genitals of a female come into contact with it. Touching the lips with aconite produces a feeling of numbness and tingling, and this was often used as a crude test for the presence of the deadly poison.

A typical list of effects for ingested aconite includes the following before death finally arrives: tingling and burning sensation in the tongue, throat and skin; restlessness; respiratory distress; muscular un-coordination; vomiting; diarrhoea; and convulsions. In Ayurvedic medicine, aconite has been particularly valuable. Generally, a popular drug down through the centuries, aconite was used in Western medicine up until the 20th Century, after which it gave way to safer and more-effective alternatives. An important application was in making liniments for muscular pain, and in toothache preparations.

The poisoner Dr George Henry Lamson first learnt of aconite when he attended a lecture by Robert Christison at Edinburgh University in the late 1870s. Lamson went on to learn that the deadly poisonous chemical extracted from the root of monkshood had no antidote, and could not be detected in the human body.

Lamson had a heroin habit that was causing him serious financial problems. It seems the habit came from his days as a doctor in the army, after he was wounded in the chest, and took morphine to ease the pain. On leaving the army and setting up his own medical business, the heroin habit turned to serious addiction. Ill health and strange behaviour followed. Eventually this led to the failure of his business and a pile of debts. He saw one way to rid himself of the spiral of debt, and that was to acquire some of the family money, which was not going to be easy, for he had already squandered the whole of his wife's money.

There was one possibility. If his wife's brother Percy died, then Percy's money would come her way and ultimately into Lamson's hands. Percy, aged 18, was boarding at Blenheim House School in Wimbledon. Lamson decided to pay the boy a visit and take him a Dundee cake. But this was no ordinary cake, for Lamson had laced its raisins with Morson's Aconitine, a brand of purified aconite extract that he had recently purchased from a London pharmacist. On arrival at the school, Lamson gave the cake to Percy, and fairly soon departed. About a half hour later the boy became violently ill and died.

It was not long before suspicions were voiced. The case was reported in the press, and the pharmacist who had sold the poison contacted the police. An investigation followed and the case was brought before the court. The expert evidence came from analytical chemist Thomas Stevenson, who performed tests on Percy's internal organs during the post mortem, that were then in the early stages of decomposition. Stevenson's tests amounted to detecting a numbing sensation when he tasted the extracts from the body's organs and body fluids. Further details of this case are given in the *Testing for Toxins* chapter.

Despite the unpleasant task, Stevenson was able to detect a tingling sensation, numbness and swelling in his throat, and noted how the numbness lasted for some hours. He compared these sensations with a sample of aconitine that was similar to the one Lamson had purchased from the pharmacist. The conclusion was that aconitine was present in Percy's urine and stomach contents. There was much dispute about the tests, with other professionals challenging Stevenson's findings. However, the outcome was that Lamson was found guilty and executed on 29th April 1882.

3.23 WHICH ONE FOR WITCHES?

Belladonna, also known as 'deadly nightshade', has purple flowers and purple-black berries. The Italian word belladonna means 'beautiful lady' and derives from the appearance of the eyes after dropping into them a minute amount of a solution prepared from the plant. The effect was used to enhance a lady's appearance by enlarging her pupils.

Used from early times as a plant preparation, it was not until the 19th Century that the active compounds, atropine, hyoscine and hyoscyamine were extracted from belladonna in their pure form. It is the atropine that is the active component in producing enlargement of the pupils. Nowadays, atropine in its purified form is used by doctors in ophthalmology for examination of the retina. Atropine is so powerful that a solution as weak as 1 part in 130 000 parts of water was found to be effective on cats' eyes. Hyoscine also finds use in medicine as a smooth muscle relaxant. It relaxes a part of the peripheral nervous system acting as an antispasmodic, and prevents excessive salivation.

Atropine was found to be particularly useful for treating whooping cough and hayfever. It has also shown promise in treating Parkinson's disease. When applied externally, atropine is used to relieve pain from lumbago and neuralgia. Atropine also played a part as an antidote to the effects of organic phosphorus nerve gases in World War II. These gases attack the nervous system by concentrating at the nerve junctions and causing the nerve to become over-active. The atropine works by blocking the action of the organic phosphorus compound that allows the nerve cells to function normally.

Belladonna extracts have been used since early times to treat muscular spasm, but it was not until the 19th Century that belladonna's individual compounds were isolated and identified.

In ancient surgery, a paste prepared from belladonna, hemlock, mandrake and henbane was applied to the skin to render the patient unconscious.

In henbane the active compounds are similar to those in belladonna, but occur in amounts that are highly variable, making preparations based upon henbane unpredictable in terms of results. Henbane preparations for taking by mouth are extremely poisonous. In fact, henbane concoctions acquired a sinister reputation and became known as 'the witches brew'. They are capable of producing some powerfully psychoactive 'magic brews' causing visual hallucinations and the sensation of flight. Despite the high risk associated with henbane, it was popular with the ancient Greeks and was documented in *Naturalis Historia* by Pliny the Elder.

The compounds in belladonna, atropine in particular, act on parts of the nervous system that inhibit the regulation of involuntary action. Early symptoms of poisoning from atropine are increased heartbeat, dryness of mouth, excessive thirst and hot, dry, red skin. In cases of more severe poisoning, the symptoms are pupil dilation, difficulty urinating, delirium and hallucinations, and excited erratic behaviour that can become dangerous. In the worst cases, convulsions and paralysis of parts of the brain led to coma and death. Children playing in areas where belladonna grows have sometimes been poisoned by atropine and related compounds.

Preparations of the belladonna plant were popular in early Roman times. Livia used it to poison her husband the Roman Emperor Augustus, and belladonna found use as a weapon to poison the water and food supplies of the Roman's enemies. Women were given belladonna preparations when they took part in bacchanalian orgies, in which they performed naked frenzied dances and threw themselves at the waiting men. Presumably it was the psychoactive power of belladonna to produce delirium and hallucination that created the effect.

Purified hyoscine was used in the early 20th Century in mental institutions as a sedative, and is in current medical use for painful menstruation and to relax the uterus during labour. Administration is by mouth injection or skin patches, and has the trade names Boscopan and Scopoderm.

Mandrake is similar to belladonna and henbane in terms of the active chemical within it. However, there is so much folklore attached to it that it has become something of a special case. Mandrake has a very short stem and thick forked fleshy root that was thought to resemble the human form. All parts of the plant are poisonous due to the presence of hyoscine and hyoscyamine. It has been used since early times for its emetic and narcotic properties, but is more widely known for its magical properties. Of all the plants associated with magic, this one stands supreme.

Folklore has it that mandrake roots absorb the dark earth spirits, and these give the roots their supernatural powers. In the Middle Ages the mandrake was associated with bodies of executed criminals and was said to grow in earth where hanged men had dripped semen. Alchemists experimenting with the herb were said to have projected human seed into the animal earth in order to ensure the growth of a mandrake.

In classical Jewish references, mandrake is said to help barren women to conceive. The Old Testament refers to it in Genesis 30: 14–22 where the mandrake is effective in helping Leah to become pregnant. And in the *Songs and Sonnets* of John Donne (1572–1631) we read:

Go catch a falling star,
Get with child a mandrake root,
Tell me, where all past years are,
Or who cleft the Devil's foot.

During the Medieval period it was believed that if the mandrake root was pulled from the ground it would shriek. All who heard its scream would be driven mad or even die if they failed to block their ears.

To remove the root safely it was recommended that a furrow was dug around the root until its lower part is exposed. One end of a cord should then be attached to the root and the other end to a dog, preferably a black dog. The dog's master should then beat it with a stick so that the dog runs off and the mandrake root is pulled from the ground. As soon as the mandrake is uprooted, the dog dies in agony. We can imagine that the collection of several roots, as would be needed to satisfy a healthy demand for magic potions, would impose some strain on the supply of black dogs.

In English folklore there were sex differences, with some texts reporting mandrakes and womandrakes. Old herbals show pictures of roots as male or female forms with bunches of leaves growing out of the head. The mandrake took on such importance that, in the 16th Century, a thriving trade developed and caused shortages in the supply of mandrakes. Counterfeit mandrake roots appeared. These were usually bryony carved into human shape, sometimes with tiny shoots of grass attached to simulate pubic hair.

In Shakespeare's *Romeo and Juliet* we note "shrieks like mandrakes torn out of the earth". And in J. K. Rowling's *Harry Potter and the Chamber of Secrets* the mandrake was cultivated for its root by Professor Sprout. Machiavelli, in his play *The Mandrake* has a plot that revolves around the use of a mandrake potion as a means to bed a woman. D. H. Lawrence mentions the weed of ill-omen when referring to the mandrake.

Medieval witches were particularly familiar with the magical qualities of mandrake (in this context often referred to as 'mandragora'). They had love potions and flying ointment. The latter, when rubbed on the skin, produced a state of dissociation, a trance, and the perception of flying to the witches' Sabbath, presumably on a broomstick with the cat sat on the twiggy bit at the back.

Mandrake certainly played a major role in the days of witchcraft and magic, but it was not alone. A range of hallucinogenic ointments was available, with deadly nightshade and henbane also on the scene as active ingredients. Atropine, from deadly nightshade, may have accounted for the hallucinations that led the witch to believe she was on a 'trip' on her broomstick to meet other witches at the Sabbath.

In 1966 a German professor (Will-Enrich Peukert) experimented with witches' brews based on belladonna and henbane. Rubbing the concoction onto his forehead, and with his colleagues doing likewise, he had an interesting experience that he later reported. Apparently he fell into a 24-hour sleep and dreamed of frenzied dancing, weird adventures and Medieval orgies.

Hemlock, also known as 'poison parsley', is found in Britain and Central Europe, and is the source of coniine, a poison found in high concentration in the plant's seeds. Coniine is a dangerous neurotoxin that disrupts the nervous system, causing paralysis of the muscles, which leads to death. The poisoning effect can be prevented if it is interrupted by artificial ventilation until the effects wear off, which takes between 48 and 72 hours.

In small doses, hemlock was used as a sedative and antispasmodic, and in early Greek and Persian medicine hemlock was used to treat arthritis. In addition to coniine, hemlock contains methylconiine and conhydrine. It is the coniine that is the powerfully active compound and which paralyses the nerves. However, purified coniine is used in modern medicine particularly for treating asthma and whooping cough.

Hemlock was popular as a suicide drink. Preparations made from the plant in ancient Greece were used for the execution of those condemned of serious crime, giving rise to the name 'Athenian state poison'. Socrates, the philosopher, was condemned to death for teaching his students radical ideas that ran contrary to those of the state. In the writings of Plato, the death of Socrates by hemlock poisoning is described, and Shakespeare's Hamlet also refers to it.

Water hemlock, sometimes confused for ordinary hemlock, resembles parsnips and has a sweet taste. Cattle are attracted to it because of the taste, with fatal results. Hence its other name, 'cowbane'. In water hemlock, the poisonous component is the chemical cicutoxin.

3.24 PUFFS OF POISON

The deliberate inhalation of tobacco smoke provides a rapid release of the psychoactive drug nicotine. In obtaining a nicotine fix, the body is also exposed to a wide range of other poisons, some of which have an acute effect and others of which are chronic poisons. The smoking of tobacco involves distillation and pyrolysis, when the plant material is heated to high temperature in the miniature furnace. The nicotine is vapourised, and cellulose is broken down, releasing poisonous formaldehyde and acrolein.

Nitrate in the cigarette paper reacts to form nitrosamines, which are carcinogenic. The toxic gas carbon monoxide is produced in the burning zone, due to incomplete combustion of carbon at high temperature. The heat also forms free radicals that enter the bloodstream, where they react with a wide range of the body's natural chemicals.

A recent development aimed at making smoking less dangerous is the electronic cigarette. This works by evaporating a solution of nicotine. As such, the smoker inhales nicotine and water vapour, and the other risky substances are avoided. Nicotine is a poisonous alkaloid found in the dried leaves of the tobacco plant. Tobacco was grown in Europe from the mid-16th Century, when seeds were brought by the French diplomat Jean Nicot (1530–1600) – hence the name 'nicotine'. In its pure form, nicotine is one of the most powerful poisons.

In addition to smoking, nicotine also finds use as an insecticide, being sold as a 40% solution of nicotine sulphate. It is a colourless liquid that turns brown on exposure to air and light. It has an acrid taste, gives a burning sensation and is efficiently absorbed through the skin. The burning sensation can be experienced when chewing nicotine chewing gum. Nicorette patches, for those trying to stop smoking, enable a slow release of nicotine into the body *via* the skin and blood vessels immediately beneath the point of attachment. Ingestion of large doses of nicotine produces the following symptoms: extreme nausea; vomiting; evacuation of bowel and bladder; mental confusion; convulsions; and death.

Count Bocarme took a particular interest in nicotine, and attended lectures by Professor Loppens, an expert in the newly developing field of alkaloid chemistry. At the time, cases of alkaloid poisoning were known, but there was no way of detecting their presence in the human body. This was especially so with nicotine, a fact that the Count became interested in. But his interest was not purely academic, for the Count had an ulterior motive, and that was to poison Gustave, the Countess's brother to whom her father had left the family estate upon his death.

Gustave was unmarried and was in a poor state of health after having a leg amputated. If he were to die, the Countess would receive the whole of her father's estate, and this would then pass to the Count and Countess, and haul them out of their financial straits. The couple decided to poison Gustave. With his poor state of health and a poison that could not be detected, they were confident that his death would pass for death by natural causes. Learning more about alkaloid chemistry from the professor enabled the Count to make plans. He planned to try out the extraction of nicotine. He purchased a retort, set up a laboratory in an outbuilding, and proceeded to distil the nicotine, a volatile oily liquid, from tobacco. This took many days, as there was much trial and error before he had found the correct procedure and the appropriate conditions.

He claimed that his laboratory was for making eau de Cologne. Success eventually came and pure nicotine was isolated. Tests were performed to examine its poisonous nature, and several dead cats were evidence of its potency. It must have seemed odd that a laboratory purported to be making perfume should be responsible for a number of animal deaths.

The time had arrived and the Count and Countess invited Gustave to dinner. Elaborate preparations were made to ensure the servants were not around at the critical moment. Towards the end of the evening Gustave collapsed and cried out in agony, the Countess announcing that he had suffered a stroke and was dead. Gustave's body was put into a spare room and the Countess set about doing a thorough cleaning job of the dining room. The servants soon became suspicious and visited the local priest, who organised for doctors to go to the house to examine the body.

The doctors noted corrosive burns around the mouth and inside it, and concluded that he'd ingested a corrosive liquid. They removed the tongue, digestive system and liver and despatched them for analysis. At the laboratory, tests were carried out. Among them were tests on dogs to see how nicotine affects the mouth. The couple were arrested, and later the Count's distillation apparatus was discovered secreted behind a boarded-up section of a wall.

3.25 SHERLOCK'S SPECIAL SPIRIT

In the early 1800s the alkaloids attracted a lot of attention, and extraction techniques were developed to separate the pure alkaloid from the crude plant material. Many of the experiments focused upon the extraction of pure morphine from opium, which comes from the poppy *Papaver sominferum*.

Morphine had recognised benefits as a medicine but knowing the appropriate dosage was a problem, as a patient could easily be poisoned by accidental overdose. An example of a medicine based upon morphine was the painkiller laudanum. It contained up to 10% of raw opium dissolved in alcohol and water, and was frequently taken by the fictional character Sherlock Holmes.

Despite most of the interest in alkaloids beginning in the 19th Century, there had been previous interest and experiments. Paracelsus in the 16th Century found that alkaloids were generally more soluble in ethanol than water. He prepared solutions of opium extract in ethanol, which we now call 'tincture of opium'. This became the basis of laudanum. Taken orally, laudanum became widely used as an analgesic, a cough medicine, and for control of diarrhoea.

The opium content of laudanum meant that the medicine contained some addictive alkaloids such as morphine and codeine. Being a powerful narcotic, it led many users into addiction and then to death by respiratory depression. It was withdrawn from general sale but is still available in some countries on prescription. Morphine preparations still play a role in controlling diarrhoea. Today we can buy a kaolin and morphine mixture from our local pharmacy. The kaolin is a finely powdered clay that can adsorb toxic chemicals and act as a bulking agent. The morphine serves to reduce water in the large intestine. Used against an attack of diarrhoea from food poisoning, it is effective, but one doubts it has quite the same punch as a dose of laudanum.

3.26 POISON PELLET

The castor oil plant has seeds with hard shells that contain one of the most powerful toxins, ricin. Ricin is toxic by inhalation or injection, but if ingested it has less effect because the digestive system breaks it down. Despite this, many children have died after chewing castor oil seeds.

In ricin poisoning, the symptoms appear up to 24 hours after exposure. Gastro-enteritis and jaundice are the early signs, and these are followed by death due to heart failure. When ricin is injected, the results compare with those of snake venom in terms of potency. For snake poisons there are often antidotes; for ricin poisoning there is no known antidote. However, ricin has a positive side in that it has antitumour properties, and is being tested for use in cancer therapy.

The umbrella murder case was thought to be poisoning by ricin. In London on 7th September 1978, Georgi Markov felt a sharp stab in his leg that he put down to a man having jabbed him with an umbrella as he walked by. Markov later experienced a high temperature and vomiting. He was taken to hospital where tests suggested a blood infection, possibly septicaemia. His fever became worse and he soon died. The authorities were suspicious and a post mortem was ordered, during which a tiny metal sphere was found embedded in his leg at the site of the umbrella jab.

Porton Down, the British government chemical defence laboratory, took on the examination of the metal sphere. They found the pellet had two small holes, each about 1.5 mm in diameter, drilled at right angles to each other and capable of containing about a 0.5 mg of liquid. Assassination was thought to be the motive, as Markov had written much damaging criticism of the communist regime in his homeland Bulgaria. When the Soviet Union fell, it was revealed that secret KGB records made reference to an umbrella device for injecting poison spheres into victims.

Twenty-five years after the assassination of Markov, six Arab men were arrested at two London addresses, where castor oil beans were found along with equipment to extract the ricin. Due to the ease of extraction and the fact that it is highly effective through inhalation, ricin is attractive to bio-terrorists.

3.27 LETHAL LEAVES

The red stalks of the rhubarb plant are a popular dish once thoroughly cooked and sweetened, but rhubarb leaves are quite different. The leaves must never be eaten, for they contain poisonous oxalic acid, and this has been the cause of many fatalities from kidney failure. The problem is that the oxalic acid can form crystals in the kidneys, and these block the filtration action that removes unwanted chemicals from the bloodstream.

Laurel leaves are also poisonous if ingested. The deep green leaves contain cyanogenic glycoside. This chemical is broken down by the digestive system to release cyanide ions. Although a healthy adult can tolerate a certain amount of this, there is a limit. Laurel leaves or the liquid that can be squeezed from them contain a large concentration of glycoside, which releases cyanide in the digestive system. The poisoner Graham Young, whilst being held at Broadmoor high-security hospital, was thought to have extracted cyanide from the leaves of a laurel bush growing in the hospital grounds.

The laburnum tree, with its lovely golden flowers, is a familiar sight in most gardens. Most parts of the tree are poisonous, particularly the seed pods, which contain high concentrations of the toxic alkaloid cytisine. I remember as a child picking these little pods from a laburnum in our garden thinking that they were miniature pea pods. I picked; I opened; I tasted and then spat out. The contents of the pods certainly didn't taste like the pea pods we bought from the greengrocer's shop.

3.28 MOULDS AND MUSHROOMS

There are around 50 000 species of fungi, of which the common examples are yeasts, rusts, moulds and mushrooms. In terms of poisoning it is the mushroom that needs most caution. The poisonous nature of many fungi is well known, and distinguishing what is safe to eat from what is poisonous is not a task to be taken lightly, and is best left to the expert.

Toxic fungi attack the liver and kidneys with fatal results. Some fungi are particularly effective in absorbing the heavy metals cadmium, lead and mercury from natural soil or contaminated soil from industrial activities. For instance, the ingestion of edible mushrooms grown on mercury-contaminated soil would result in mercury poisoning.

Fungi are living organisms often thought of as belonging to the plant kingdom. However, they do not carry out photosynthesis but feed upon the organic matter of other living systems, dead or alive. In this respect it seems that fungi are more akin to animals than to plants.

To many of us, fungi have a sinister quality. They appear in the forest in the dark of the night and thrive on dead tissue. In the words of well-known mycologist John Ramsbottom in his book *Poisonous Fungi*, 1945: "To many the very word fungus suggests something mysterious, something morbid. Some of the early herbalists seeking a derivation of the word found it in funus (a funeral) and ago (to put in motion)."

Fungi break down their sources of food using digestive enzymes, and those same enzymes, if ingested by humans, can also break down the cells of major organs such as the liver. Many fungi are toxic because of their enzymes as well as the other poisonous chemicals, known as 'mycotoxins', which are present in their tissue. In general, the chemicals are a kind of alkaloid known as 'quinonoids'. In fruiting fungi such as mushrooms there are psychoactive chemicals used for recreation or in spiritual ceremonies. However, ingestion of these and preparations made from them does not come without risk.

Most fungi are not toxic, and we make good use of them. We eat mushrooms and truffles; we use yeasts in fermentation to make the bread dough rise and to produce the alcohol in wine; we use fungi as a source of antibiotics; and in agriculture, fungi are used to kill insect pests. Ice man, the 5300-year-old corpse found in the Austrian Alps carried two species of mushroom. The amount was insufficient as a food source, so what were they for? Perhaps they had a medicinal use.

3.29 DEATH CAP

This can grow in a few hours or up to three days, and is responsible for 95% of all mushroom poisonings. Its main toxin is amanitin, and as little as 50 g of mushrooms contain enough of the poison to kill an adult. There is often the thought that thorough cooking of mushrooms destroys the toxins. This is not the case with Death Cap mushroom, for the toxins are resistant to heat.

The Death Cap (*Amanita phalloides*) has been responsible for many deaths. It is thought the Roman Emperor Claudius was poisoned by the mushroom in 54 AD and that the Holy Roman Emperor Charles VI was killed by Death Cap in 1740. Recently, refugees arriving in Germany from Syria were poisoned by Death Cap mushrooms while camping in the woods and foraging for food.

In 2013, 57-year-old Christine Hale from Bridgwater, Britain died of major organ failure four days after eating soup made from Death Cap mushrooms. She had apparently picked them from beneath a nearby tree in the belief that they were edible. With many toxic fungi the bad taste acts as an early warning, but Death Cap are sweet and have a pleasant taste.

There is, as yet, no way of neutralising the poisons, and the only chance of surviving is by having a liver transplant. The first symptoms arrive just 10 to 16 hours after ingestion, and include abdominal pain, vomiting and diarrhoea. After these have passed, the patient seems to have recovered, but the toxins are still there and working away attacking the liver, heart and kidneys. The toxins destroy the liver function by inhibiting enzymes in the liver and blocking protein synthesis. However, if the patient receives prompt medical attention after ingesting the mushrooms and is given a large dose of penicillin and vitamin C, there is a chance of surviving.

3.30 MUSHROOM METALS

Even non-toxic mushrooms can be poisonous when they accumulate cadmium, mercury, lead, copper and chromium. Absorption of some or all of these poisonous metals is known to occur when mushrooms have grown on land contaminated from industrial use.

Growth of mushrooms on old landfills, tips, fly ash lagoons at coal-burning power stations, and spoil from mining activities and mineral processing also poses a risk. Dumped fly ash from coal-burning power stations leaves large areas of ground contaminated with the oxides of heavy metals, and this poses a risk if mushrooms have grown nearby. Mushrooms growing at the side of the road may have accumulations of lead that has remained in the soil from the days when cars ran on leaded petrol.

Water draining from any of these may carry poisons to seemingly clean ground. Radioactive isotopes of metals such as caesium-137, from nuclear explosions and power station disasters such as Chernobyl, have been found to build up in mushrooms.

3.31 ST ANTHONY'S FIRE

This is a parasitic fungus that grows on rye and wheat. It produces some important alkaloids such as ergotamine, which is used for the treatment of migraine. Eating bread made with rye infected with the fungus has resulted in ergotism, in which the symptoms are vomiting, diarrhoea, headache and, more seriously, gangrene of fingers and toes.

Ergotism caused 11 000 deaths in Russia in 1926 when infected rye was used. In Medieval times ergotism was known as St Anthony's Fire because of the inflamed appearance of the gangrenous tissue. It was believed a visit to St Anthony's tomb was the cure. The alkaloids in ergot are important in medicine and are now used to treat migraine and sex disorders. One of these alkaloids is ergotamine and is related to the hallucinogenic drug LSD (lysergic acid diethylamide).

3.32 PENICILLIN

There are some serious poisons in the world of fungi, and many human deaths have resulted. However, it is worth noting that penicillin mould produces a toxin that can be used to protect us from certain bacteria. The material in question is penicillin. In fact, there are several penicillins that now form the basis of compounds we take in small doses for their antibiotic action. Familiar examples are amoxicillin and ampicillin. In effect we dose ourselves with a fungal poison to prevent us from being dosed with bacterial poison.

Penicillin's action of bacteria was discovered by the Scotsman Alexander Fleming whilst carrying out research at the medical research centre at St Mary's Hospital in London. He had been carrying out research to find an antibacterial agent that could cure sepsis in open wounds such as he had seen during his time in the army during the First World War.

He noticed some petri dishes that had been put to one side for washing but forgotten about for some days. It appeared that the cultured bacteria had areas where the bacteria seemed to have been killed. Further work revealed that the poison produced by a mould was responsible for the bacteria being destroyed.

Fleming grew cultures of the moulds, and showed that they were effective against many types of disease-causing bacteria. The mould was identified as being penicillium rubrum. The World's first antibiotic had been discovered, but it proved difficult to isolate and produce in useful amounts. Years later, with funding from the American and British governments, it was produced in sufficient quantities to be available for treating bacterial infections. It became available in 1941 (Table 3.3).

Table 3.3 Poisonous fungi.

Fungus	Poison/delivery/symptoms/antidote
Aflatoxin fungus (*Aspergillus flavus*)	Tumour toxin may cause cancer of liver in humans. Grows in warm humid conditions where nuts and cereals are stored.
Fools webcap (*Cortinarius orellanus*) Deadly webcap (*Cortinarius rubellas*)	Poland. 1950. Over 100 died after eating Fools webcap.
Panther cap (*Aminata pantherina*)	Causes hallucinations and sickness. In appearance resembles the blusher mushroom, which is edible. Contains muscamol.
Fly agaric (*Aminata muscaria*)	Red–white spots, fairy tale toadstool, small amounts cause sickness and delirium. Larger amounts are fatal. Muscin is the toxin, which may be neutralised with atropine. Symptoms: hallucinations, nausea, diarrhoea, convulsions and death.
Fool's funnel (*Clitocybe rivulosa*)	The sweating mushroom. Causes nausea, diarrhoea, blurred vision, and sweating. Can be fatal. Contains muscamol.
Jack O' Lantern (*Omphalotus olearius*)	North America. Glows in dark due to luminescent chemicals. Symptoms, cramps, diarrhoea.
Yellow stainer (*Agaricus xanthodermus*)	Causes sweating, flushing and severe intestinal problems but does not affect all in same way. Once regarded as edible.
Destroying Angel (*Amanita virosa*)	Produces amanotoxins that act on the major organs. No antidote. Only chance of survival is liver transplant.

CHAPTER 4

Mineral Matters

Here we take a look at poisons that are neither animal nor vegetable but are mineral. They are to be found in the atmosphere, in the oceans and on, or beneath, the ground. On our planet there are a vast number of minerals, some of which are highly poisonous but most are deep beneath the ground and out of harm's way. In addition to the poisons that are present on Earth, we also have extra-terrestrial materials arriving on Earth each and every day.

4.1 GHASTLY GASES

Nature produces a range of poisonous gases, and man's activities add to this. Waste gases enter the atmosphere from industrial discharges, some of which are legal, some illegal, and some not known about. We also have increasing amounts of waste gases from road vehicles and aircraft. In our cities the air is poisoned by nitrogen dioxide from diesel vehicles. There is growing concern about the damage to the lungs through inhalation of this gas.

Unburnt hydrocarbons emitted from cars react in sunlight to convert oxygen into ozone. This forms ground-level ozone, which is another poisonous addition to the air we breathe. Also, from our engines we have the well-known carbon monoxide and traces

---

Poisons and Poisonings: Death by Stealth
By Tony Hargreaves

Published by the Royal Society of Chemistry, www.rsc.org

of sulphur dioxide and hydrogen sulphide. The main gases from our internal combustion engines are carbon dioxide and water vapour.

The combustion of 1 L of petrol produces water vapour that condenses in the cold air to give over 1 L of rain water. Perhaps we should reflect upon this when complaining about the increasing volume of rain. For instance, we might question just how much 'new water' is put into the atmosphere from a transatlantic flight that consumes over a 100 000 L of kerosene on each single flight, and then there's the return flight.

Modern agriculture with its huge populations of sheep and cows makes a contribution to atmospheric poisoning. The animals release methane into the air from their digestive systems. This might seem somewhat trivial or amusing, but consider how much time the cows and sheep spend eating, belching and breaking wind. The volumes of gas involved are worrying, as methane has a higher rating as a greenhouse gas than carbon dioxide.

Methane emission from man's activities is increasing and adding to global warming. The main gas given off when organic waste decays is methane. With our enthusiasm for creating waste and dumping thousands of tons of food into landfill we are adding to the methane emissions. Some might point out that much of the methane from landfills is burnt off. This simply converts the problem to one that is less serious, as the methane then produces more carbon dioxide.

Methane from decomposing organic matter has both natural and man-made origins. Watercourses are often used for dumping waste. In the days of the industrial revolution all manner of filth entered the rivers and canals. In one Yorkshire mill town the canal basin contained so much waste that, when a lit candle was placed near to the surface, the water caught fire. That was due to methane evolved from decaying matter. This is something that also happens in marshes and swamps. This natural emission is known as 'will-o'-the wisp' (*ignis fatuus*) and is also caused by methane, but how it self-ignites is a mystery.

4.2 UNDERGROUND GASES

Gases are also produced in the ground from the waste of human activities. There are brownfield sites and landfill sites containing industrial and domestic waste that was dumped in the past. Such waste may produce toxic and flammable gases or vapours that diffuse through the ground depending upon the local geology.

The Earth's surface has a huge capacity for releasing dangerous gases from beneath the ground. From a volcanic eruption there is sulphur dioxide, carbon monoxide and heavy metals such as mercury. At the bottom of some oceans there are vast accumulations of methane clathrates, which if disturbed erupt and deliver enormous amounts of methane.

4.3 ROTTEN-EGG GAS

Carbon dioxide is a colourless, odourless gas that is denser than air. One of its main uses is in extinguishing fires, and it can also extinguish life. Carbon dioxide is inhaled from the air we breathe in which it is present at a concentration of about 400 ppm. Before industrialisation and the burning of fossil carbon the level was about 250 ppm. As it is a waste product of our respiration, we would not think of it as a poison, but, when the concentration is high enough, it can kill. The way in which it kills is that, as the proportion of in the air increases, the proportion of oxygen and other gases decreases.

A level is reached at which there is insufficient oxygen for the body's essential respiration, and death results. We would normally tend to think of this more as a case of suffocation and oxygen starvation. However, it is still a form of poisoning. A recent case was when a truckload of illegal immigrants from China arrived in Britain. They had suffocated because at some stage in their journey someone had closed the ventilator accessible only from the outside. There are suggestions that this was done deliberately, which may make the case one of mass murder.

Carbon dioxide poisoning occurred on a massive scale in the area surrounding Lake Nyos in Cameroon. The lake sits in the crater of an inactive volcano. Carbon dioxide released from the magma built up in high concentration at the bottom of the lake. However, the carbon dioxide did not simply bubble to the surface as we might expect for a chemical that is normally a gas. The carbon dioxide was under a considerable depth of water that kept it under high pressure, since each metre depth of water produces a pressure of 1 ton m^{-2}. This caused the carbon dioxide to be liquid until some action, such as a landslide or the rumblings of volcanic activity, stirred it up and destabilised it.

This occurred in 1986 when millions of tonnes of carbon dioxide erupted and made its way to the surface as a gas. Being heavier than air, it formed a dense blanket over the water and then spread to the nearby villages as the eruption continued. The cloud of dense gas killed 1700 people and 3500 livestock as it spread in this large-scale poisoning by nature. This put the lake in the Guinness Book of Records as being the deadliest lake.

The primary effect was that of the carbon dioxide, but the problem was made worse by a secondary effect due to the deadly poisonous gas, hydrogen sulphide. Some of the surviving villagers reported the smell of bad eggs. The smell is a characteristic of hydrogen sulphide and the facts are consistent with this. Hydrogen sulphide gas, which is more poisonous than prussic acid gas, is produced when metal sulphides from volcanic rock react with carbonic acid in the water. Clearly, with the sulphide rocks and high concentrations of acidic water, all that was needed was for the lethal concoction to be given a stir.

An interesting aspect of hydrogen sulphide is that in low concentrations it smells of rotten eggs but in high concentration, and at lethal levels, it has no odour because it desensitizes the olfactory nerves. The person who was inhaling the gas is fooled into thinking that it has gone away as there is no more smell. Needless to say, many people have been unknowingly poisoned by hydrogen sulphide. Most of our experience of hydrogen sulphide is from the breakdown of amino acids containing sulphur that occurs in the human gut. When a person breaks wind, hydrogen sulphide is the smelly gas.

4.4 METALS

There are about a hundred chemical elements, and most of them are metals. Many are present in only minute amounts in the Earth's crust and are of little importance in everyday life. Metals usually occur as ores, which are compounds such as oxides and sulphides. Some metals such as gold are so unreactive that they are found un-combined with other elements.

Iron is occasionally also found un-combined in bits of asteroids that have collided with the Earth over millions of years.

At the Royal Observatory in London there is a part of an asteroid that fragmented to form bits that collided with the Earth. It is known as the Gibeon Meteorite, and landed in Namibia in prehistoric times. The estimated mass of all the fragments amounts to 15 000 kg. Chemical analysis shows it to be 92% iron, 7% nickel, and small amounts of cobalt, iridium, germanium and phosphorus.

Gold, on the other hand, forms very few compounds and generally does not become bio-available in our body chemistry. Platinum is slightly more reactive and can form a chemical that is reactive in the human body. The chemical here is cisplatin and is used in chemotherapy as it may inhibit the growth of a tumour.

Precious metals like gold and platinum may not be significant in terms of human poisoning, but there are many other metals. Going back through the history of poisoning we find a small group of regulars: arsenic; antimony; mercury; and lead.

Caution must be exercised in how we describe poisons based upon metals. Simply to state that a person was poisoned with arsenic is not precise enough for a scientific study of poisons. What form did the arsenic take? Was it the grey metal? Was it white arsenic? Was it an organic arsenic compound such as cacodyl? Unfortunately, white arsenic is, by tradition, simply referred to as 'arsenic' in much of the literature on poisons.

The different forms of arsenic have different poisoning powers with, surprisingly, the pure metal being the least toxic. In some poisonings the chemical to which the person was exposed was not what actually killed them. There are instances in which a non-toxic chemical was absorbed into the body and then it reacted and changed into a toxic chemical.

4.5 ARSENIC

The Bronze Age owes much to arsenic, for although bronze is usually thought of as an alloy of tin and copper, it can also be made with arsenic and copper. It is now believed that, during those early years of metallurgy, arsenic poisoning was a problem. Smelting copper ore in making the bronze would have, from some sources of ore, released arsenic vapour, which was no doubt inhaled with fatal results. Later, in Greece, the god of blacksmiths and metallurgy (*Hephaestus*) was shown with deformed feet possibly due to arsenic poisoning.

Arsenic is naturally present in trace amounts in the human body. Many arsenic compounds produce a garlic-like odour that can be a clue to ingestion. Arsenic as the element is used in making alloys, whereas arsenic compounds have been used in tonics, weedkillers, cosmetics, rodent poison, fly papers (Figure 4.1), pigments such as Scheele's green, glass and enamels.

Gaseous arsenic chemicals are treacherously poisonous and include arsine and cacodyl. Arsenic's history goes back to antiquity, when its compounds were used in everyday life for legitimate purposes. The chemicals found little use as poisons because they all had taste, smell and colour, which made them easily detected if slipped into food or drink.

Alchemists eventually discovered how to prepare pure white arsenic, a tasteless and odourless white powder. What more could the would-be poisoner ask for? The compound was readily available, it was lethal in small amounts, and it could not be detected in food by taste or smell.

Figure 4.1 Flies were a menace in the filthy conditions that many people had to live in during the Industrial Revolution. A common means of killing them was to use fly papers, which were usually suspended from the ceiling of the affected room. The papers were impregnated with arsenic that was easy to extract. Florence Maybrick poisoned her husband with arsenic that she had extracted from fly papers. She was convicted of murder in 1889.
© Shutterstock.

Those who carefully planned a poisoning by means of arsenic would often go for chronic poisoning. Despite this being a long-term effort, it was likely to raise no suspicion. For example, a young husband suffering from what appeared to be gastric fever that got worse month after month was unlikely to invite questions. When the fever eventually killed him it would, more likely than not, be recorded as death by natural causes. However, if that young husband had suddenly dropped dead it would be a very different matter and curiosity would be aroused.

It is no surprise that arsenic poisoning enjoyed widespread popularity down the ages, especially in Italy during the 16th and 17th Centuries and again in the 19th Century in what became something of a golden age for poisoners. In France, arsenic became known as *poudre de succession*, 'inheritance powder'.

White arsenic was widely used for killing insect pests and rodents, and it was a common component of many cosmetics. Examples of products advertised were Dr MacKensie's improved arsenic wafers, which claimed to produce a most lovely complexion. There was also a brand of arsenical soap on sale for 1s 3d a tablet. In the 19th and early-20th Centuries it could be purchased from the local pharmacy. The purchasing of white arsenic in large amounts, or in small amounts on a frequent basis, would of course have been suspicious. It would be wise not to go to the same pharmacy too often.

Consider the scenario involving a young woman going to the pharmacist to buy arsenic. The pharmacist asks, "And what are you going to use the arsenic for?" The young woman replies, "To poison my husband". Clearly, she would leave empty handed. She goes to another pharmacist and he asks, "What will you be using it for?" This time the young lady replies, "To poison a rat". She succeeds and comes away with a bag of white arsenic.

However, buying fly papers, which contained arsenic, was much easier as there were no awkward questions. In the murder of James Maybrick by his wife Florence, fly papers were used. The fly papers contained sodium arsenite, a highly soluble form of arsenic and therefore extremely poisonous. In making the papers, brown colouring was added along with a bitter-tasting substance to reduce the chances of poisoning.

Florence prepared her poison simply by soaking the papers in water for a few hours and pouring off the brown liquor produced.

As the liquid had both bitter taste and brown colour, it needed to be administered in a drink that would also have similar properties if it was not to be noticed. To achieve this, Florence put her poisonous liquid into James' tea and in his arrowroot. In 1889 James died and Florence was under suspicion.

A search of her room produced evidence of the fly paper extraction along with another source, which was black arsenic. It appeared that she had set up to extract white arsenic from black arsenic. Black arsenic, available from the pharmacy, was in fact white arsenic mixed with soot in order to reduce the likelihood of it being used as a poison.

Florence had clearly done her chemistry homework and learnt how to dissolve the arsenic out of the black powder and remove the insoluble soot using a handkerchief as a filter paper. Specimens of her filtrate, which was almost colourless, were there as evidence.

As a part of Florence's defence she claimed that the arsenic and equipment were being used for experiments in the preparation of cosmetics. There may be a grain of truth in this as after James' death Florence still had her chemistry set with raw material and product. In those times white arsenic was often applied as face powder to improve complexion.

Post-mortem analysis of James' liver, intestines and kidneys showed significant levels of arsenic, but these may have been a result of James' self-medication with Fowler's Solution to cure his failing sexual performance. Florence stood trial and was sentenced to life imprisonment, but she was released in 1904 after serving 15 years.

Another poisoner who was alleged to have extracted poison from fly papers was Frederick Seddon, known as the meanest murderer of all time. He sought out every opportunity to make money. The house in which he lived with his wife Margaret had ample space, and to gain extra cash they decided to rent out a room. Eliza Barrow, a 49-year-old spinster, moved in along with her 10-year-old nephew. Seddon won the confidence of Barrow who had a large amount of money in investments.

Within only a year from becoming a lodger in the Seddons' London home, Barrow had signed over all her assets to Seddon in exchange for an annuity. Being an insurance company official no doubt gave her a degree of reassurance. It was early

September 1911 when Barrow began to suffer agonising stomach pains and diarrhoea. The doctor was called. She continued to get worse and decided, or was persuaded, to make a will. Seddon apparently assisted her in the preparation of the will. Eliza Barrow suffered for a further two weeks before dying.

The doctor saw her but he had other urgent business and simply issued a death certificate, omitting to carry out a proper examination of the body. Without delay, Seddon saw an undertaker and arranged for Barrow to be interred in a pauper's grave. This was completed only two days after her death. Barrow's relatives became suspicious that a wealthy woman had died and seemingly all her investments had disappeared. In November her body was exhumed and tests were performed. The police believed arsenic was the poison and that Seddon had extracted from the threepenny (3d) fly papers that he had recently been purchased.

Chemical tests were carried out on Eliza Barrow's remains but the results were not convincing and there was insufficient evidence to convict Seddon. However, during his trial at the Old Bailey in March 1912 he showed little concern for the death and adopted an arrogant and condescending manner. If he had not decided to testify on his own behalf it is likely he would have been acquitted. But he was found guilty and his wife was acquitted. Frederick Seddon was hanged at Pentonville Prison in April 1912.

Many arsenic compounds were used to good effect in the treatment of infections. In his medical research Paul Ehrlich carried out extensive work to find the best arsenic compounds for killing bacteria, especially the bacterium responsible for syphilis. In 1909, he tested his 606th compound and found that it destroyed the syphilis bacterium. This compound was arsphenamine, later to be known commercially as 'Salvarsan'.

Since then, the number of arsenic-based medicines increased dramatically but arsenic compounds have now fallen from favour as more effective and safer drugs have become available. White arsenic in the form of Fowler's Solution was used up until the 1950s. The tonic, which claimed to cure all manner of afflictions, was widely used, particularly by men. Charles Darwin took it regularly to treat a hand tremor and it is thought that he suffered from arsenic poisoning as a result.

Symptoms of acute arsenic poisoning are severe irritation of the gastro-intestinal tract, nausea, cramps, vomiting, diarrhoea and finally death. Chronic symptoms include exfoliation of skin and breakdown of liver and kidney function. There is now strong evidence that long-term exposure to arsenic compounds causes cancer. Ingestion of arsenic compounds may be, prior to death, detected by a garlic-like smell on the victim's breath, and the victim may report a strong metallic taste.

Arsenic poisoning at the sub-lethal level is often followed by a sequence of distressing events until the body has removed the poison. Much of the arsenic is removed in the urine as a soluble chemical, but there is always some left in the body where it accumulates in the hair and nails. Tell-tale marks in the victim's fingernails, known as Aldrich Mees Lines, occur with poisoning by arsenic and other heavy metals.

Many names in the history of murder are those of poisoners who chose white arsenic as their weapon. Until the means became available to detect and measure the amount of arsenic in a corpse, there must have been many more poisoners who escaped detection.

Even in more modern times, with reliable means of detecting and measuring arsenic in a dead body, the presence of arsenic, especially in an exhumed body, is not always evidence of intentional poisoning. This is because arsenic is naturally present in the human body, and buried bodies can absorb arsenic from the soil.

A famous case of accidental poisoning by arsenic was the Bradford Poisoning of 1858 in which 20 people died and over 200 people were made seriously ill after eating peppermint sweets from the local market. The manufacturer of these, in order to use less sugar and make a cheap product, used plaster of Paris as an ingredient, but on one occasion the supplier of this material mistakenly supplied white arsenic. Apparently there was some confusion over labelling and the white arsenic looked exactly like the plaster of Paris (Figure 4.2).

But the Bradford Poisoning was not the only case in which children were unwittingly poisoned by arsenic. Many a child's bedroom had wallpaper on which a pattern was printed and in which the pigment that provided the green colour was an arsenic compound known as 'Scheele's green'.

THE GREAT LOZENGE-MAKER.

A Hint to Paterfamilias.

Figure 4.2 White arsenic was accidentally put into peppermint sweets, which then went on sale in Bradford market in 1858. Twenty people died and over 200 were made seriously ill. The manufacturer of the sweets normally added plaster of Paris in order to use less sugar and make a cheap product. Unfortunately, on one occasion the supplier sold the sweet maker some rat poison, white arsenic. Apparently there was some confusion over labelling and the white arsenic looked exactly like the plaster of Paris.

Unfortunately, in damp conditions this pigment reacted with the moisture from the air and released the deadly poisonous gases cacodyl and arsine that a sleeping child would inhale over many hours. It was not until the work of Frederick Challener in the 1940s that the gas was identified, after which the green pigment was phased out in favour of lead-based pigments.

4.6 NAPOLEONIC TONIC

An interesting example of the difficulty in distinguishing accidental from intentional arsenic poisoning is the circumstances surrounding the death of Napoleon. For many years he had suffered stomach pains, and some say this is why he is often portrayed with his right hand tucked into his tunic as if he's pressing on his stomach.

Following his defeat at the Battle of Waterloo in 1815 Napoleon lived in exile at Longwood House on St Helena and died there in 1821. During this time his stomach complaints worsened, showing all the signs of food poisoning or even stomach cancer. Upon his death he was buried in St Helena where he remained for some 20 years until his body was exhumed and taken to Paris for reburial.

Ever since his demise there were suspicions that the English had hired an assassin to poison him. During the exhumation it was noticed that the body was in a good state of preservation, a fact consistent with the presence of arsenic, which acts as a preservative because it kills the bacteria that would otherwise decompose the body. The poisoning theory gained momentum and many more joined the list of those who supported the hired-assassin theory.

A closer examination of Napoleon's life, especially his days in exile at Longwood House, reveals more information about arsenic. For a start it was known that he had taken arsenic tonic for many years and that chronic ingestion of arsenic was one explanation as it often led to stomach cancer. But there was another source of exposure to arsenic. In his room at Longwood House the wallpaper was one with a green floral pattern and the green pigment was Scheele's green. The house was said to suffer from damp for much of the time, which must have meant some of the arsenic pigment releasing the toxic gases that Napoleon would have inhaled.

Clearly there are a few possibilities for Napoleon's death. A sample of his hair was recently analysed by modern scientific instruments and the presence of large amounts of arsenic confirmed. But did that arsenic come from his reliance upon tonics? Was it the wallpaper? Or, was someone spiking his food with arsenic?

Another tonic containing arsenic is to be found in live oysters. These shellfish contain large amounts, as they absorb it from the sea water and it builds up in their flesh. As such, oyster eaters have raised levels of dietary arsenic. This may be beneficial if what they say about Casanova is true. Arsenic is chemically related to zinc and behaves much like zinc in many reactions. And zinc is essential to sexually active males to keep up their sperm production.

Apart from shellfish there are other foods that contain significant levels of arsenic. Some plants have the capacity to accumulate arsenic after absorbing it from natural minerals in the soil. Rice is especially good at concentrating arsenic. There is currently concern about arsenic in food because the amount is increasing. As we pump more pollutants into the atmosphere, more ends up in the food chain and destined to become a part of our flesh.

In accounts relating to the ingestion of arsenic, we come across the arsenic eaters of Styria (Austria) who were said to eat white arsenic, but the evidence is weak. There are suggestions that these arsenic eaters were doing this as a sales promotion. They were selling the idea of arsenic being beneficial to health. The truth is more likely to relate to the fact that Styria was an area where arsenic was in abundance. Any arsenic eaten would likely have been mixed with innocuous white powders to reduce the risk of death.

4.7 PERPETUAL PILLS

The metal antimony is not normally present in the human body. It is similar in many respects to arsenic and has been known since early times, when it was obtained from natural deposits of the mineral stibnite, which contains antimony sulphide. When ground to a fine powder, it was used as a pigment, known as 'kohl', for making cosmetics. Blending the kohl with fats and waxes produced an eye cosmetic suitable for darkening eyebrows and eyelashes, and for outlining. In addition to its cosmetic properties, it found wide application in medicines, and this is probably when its poisonous properties were discovered. Pliny the Elder noted some of its medical properties.

In early Roman times, special goblets known as 'emetic cups' were made from antimony metal and were popular devices for relieving the after-effects of a night's gorging on food and wine. At the start of the celebrations the emetic cups were filled with wine and put to one side to be drunk later. The little goblets were made of pure antimony metal, and with the wine left in them for some hours, a small amount of the antimony dissolved in the wine's tartaric acid to form antimony tartrate solution.

When this was drunk, the dissolved antimony set to work and demonstrated its powerful emetic properties. By this means, the excessive amounts of food and drink were vomited. This made room for more indulgence. Thus, the celebrations could continue interrupted occasionally by a vomiting session.

In Medieval times, antimony in its metallic form also played a part as an emetic. The metal was made into small balls known as 'perpetual pills' that were used to induce vomiting. When the pill was swallowed and reached the stomach, a small amount of antimony was dissolved from the pill's surface by the hydrochloric acid in the stomach. Each pill was a massive dose of antimony and sufficient to kill many people, but when swallowed only a tiny amount of antimony dissolved from the surface of the metal.

As the pill passed along the digestive tract it left a dose of antimony that acted as an emetic. By the time the person vomited, the pill was making its way through the intestines and appeared during the next evacuation of the bowel. The pill was collected, washed and kept for re-use. As only a tiny amount of the metal pill was dissolved in the digestive system, the pill was used many times over and, in some cases, was handed down from one generation to the next. Of course if the digestive system was not in good working order the pill could have stayed too long, resulting in a large release of antimony, maybe reaching lethal concentrations.

Despite antimony being similar to arsenic, and being about as poisonous, it is less likely to cause poisoning than arsenic when swallowed. Unlike arsenic, antimony invariably causes vomiting, which effectively removes the swallowed poison. However, once distributed throughout the body, antimony stays there longer than arsenic.

During the 17th Century antimony became a popular medicine, and many hundreds of formulations were available. One of them, known as 'Earl of Warwick's powder', was in big demand when it was shown to have cured King Louis XIV of France of typhoid. Its active ingredients were antimony oxide and cream of tartar. Naturally, great care was taken to ensure that the dosage was sufficient to kill the bacterium but not enough to kill the King.

Tartar emetic, antimony potassium tartrate, became available for purchase in the mid-17th Century and remained a popular household remedy until quite recently. When dissolved in water, this chemical releases antimony and tartaric acid. It is the antimony that is the reactive part and produces the emetic action. In the correct dosage the antimony irritates the nerves in the stomach, and vomiting results. Doses larger than this can be fatal.

With such a ready availability of antimony from the local apothecary, it is hardly surprising that it was widely used for poisoning. This became a worry to the authorities and steps were taken to ban its use, but it was so popular that it remained as a medicine. It was in the early 20th Century when the chemistry of antimony was researched, and when new chemicals became available. Two of these were shown to be effective in the treatment of tropical diseases, and were named 'stibophen' and 'stilbamidine'.

Today, technology makes use of antimony and its compounds. Flame retardants based upon antimony compounds are added to some plastics, children's clothes, toys and aircraft seats. Antimony potassium tartrate is used as a mordant in textile dyeing. Antimony is added to other metals such as lead to harden it by forming an alloy. In electrical technology, antimony is used in diodes and infrared detectors. Environmental pollution by antimony from human activities, such as coal burning, has shown itself in the soil where levels of antimony are much higher than in pre-industrial times.

The most poisonous compound of antimony is stibine, which is a gas. There was, a few years ago, much concern about stibine being released from babies' mattresses. It was thought that the antimony-based flame retardants put into the mattresses could be changed by bacteria into stibine. There was much ado, but it now seems to have been a false alarm. However, stibine is of concern because when antimony-containing alloys come into contact with acid or with hydrogen, stibine can be produced. Despite stibine being a deadly poison, it has in fact been used as a fumigating agent.

With antimony being in industrial use and some of it inevitably finding its way into the environment, there needs to be monitoring. The World Health Organisation gives 0.02 mg L^{-1} as the maximum allowable concentration for drinking water. Many of the symptoms of antimony poisoning are similar to those produced by arsenic because antimony is chemically similar. This similarity means that Marsh's arsenic test can also be used for antimony. In small doses, antimony produces headaches, dizziness and depression, whereas large doses produce frequent vomiting leading to death in a few days.

The composer Mozart died in 1791 with the cause of death recorded as miliary fever. This condition produces a rash with lesions that look like millet seed and is accompanied by a raging fever. There have been several theories on his death. Some believed that Mozart, being something of a hypochondriac, was forever taking medicine. As much of that medicine was antimony-based, he likely accidentally poisoned himself. We now know that some people are especially sensitive to antimony while others are highly resistant to it. It may be that Mozart was taking the prescribed dosage but he was susceptible to it. Another theory is that Franz Hofdemel, a brother in the Freemason's that Mozart was a member of, murdered him.

The motive, it would seem, was that the composer had revealed Masonic secrets in his opera The Magic Flute. To make matters worse, Mozart had an affair with Hofdemel's wife. The day after Mozart's death, Hofdemel committed suicide by cutting his throat. There were others who were envious of Mozart and who would have benefited from his death.

In 1860 Dr Edward Pritchard married Mary Jane Taylor. Soon after, he set up a medical practice in Bridlington, Yorkshire. He then ran into debt but Mary Jane's parents helped out and the couple moved to a new practice in Glasgow where they were able to afford a live-in maid. In 1863 the maid died when she was burnt to death in her bed. It was suspected that she had been heavily drugged and was therefore unaware of the fire.

It was not long before a new maid, Mary McLeod, arrived. Pritchard soon fell in love with her. Having now acquired a mistress, he decided it was time to dispose of his wife, and in 1864 he purchased some tartar emetic from the pharmacy, which he subsequently fed to his wife over the following months. The antimony started to take effect as a chronic poison and soon Mary Jane became ill. Her mother, Mrs Taylor, came to look after her, during which time Pritchard decided she also must die. Mrs Taylor died in March 1865 and her daughter died some days later.

Suspicion was aroused, but it was not until the authorities received an anonymous letter accusing Pritchard of murder that steps were taken. Post-mortem examination of Mary Jane's body showed large amounts of antimony. When Mrs Taylor's body was exhumed, it also contained high levels of antimony. Pritchard was arrested, tried and was hanged in Glasgow in front of a crowd of many thousands. Also becoming active about this time was Dr William Palmer using antimony, cyanide and strychnine for poisonings. Palmer's case is discussed later.

Now a Polish poisoner came on the scene. He was Sverin Ktosowski and had started on a course at Imperial College, London to become a junior surgeon but dropped out and went into the hairdressing business. He met Lucy Baderski who moved in with him, became his wife and produced a baby. Things did not work out and Ktosowski left Lucy and lived with Annie Chapman for a while until she became pregnant, whereupon he left her and refused to have anything to do with her. However, he liked her surname enough to then call himself George Chapman.

The next woman in his life was Mary Spink. Together they bought a hairdressing business in Hastings and it was not long before he met Alice Penfold. He decided to remove Mary from his life and planned to poison her. He bought one ounce of tartar emetic from the local pharmacy and gave Mary a dose of it in her brandy, after which she became seriously ill. As she grew weaker, a doctor was called and tuberculosis was diagnosed. Mary died and was buried in a deep grave, eventually to have seven other coffins stacked on top of hers.

A few weeks after Mary's death, George Chapman started a relationship with Bessie Taylor but soon he was also dosing her with tartar emetic. Bessie suffered vomiting, diarrhoea and stomach cramps. She was seen by several doctors and each one provided a different diagnosis.

Bessie's mother came to look after her. This made it difficult for Chapman and his routine of constantly giving Bessie a dose of the poison was interrupted. He decided therefore to give Bessie one massive dose of poison to finish her off. This he did and soon she was dead, the cause of the death being recorded by the doctor as intestinal obstruction.

Chapman then became engaged to 18-year-old Maud Marsh and they married, but he was not satisfied, and a Florence Rayner came on the scene. Soon Maud was taken ill with conditions not unlike those suffered by his previous partners. Maud's mother came to nurse her and became suspicious of the brandy that Chapman kept giving to the girl. Soon Maud was also dead, but the doctor refused to issue a death certificate and decided upon a private post mortem. Then a further post mortem took place at the Clinical Research Association. Their analysis showed that Maud's vital organs contained almost 0.5 g of antimony.

This led to Mary's grave being opened, wherein her body was found to be in an excellent state of preservation, while the other bodies in the grave were putrid and decomposed, as would have been expected. Her internal organs were analysed and a total of 0.1 g of antimony found. Bessie's body was also exhumed and, again, a remarkable state of preservation and no odour of putrefaction was noted. Analysis for antimony showed over 0.5 g in her internal organs. It was found that the bowel had a large concentration of antimony, suggesting that the poison was administered by enema.

Chapman was arrested and sent for trial. It concluded on 18th March 1903 finding him guilty of the murders of Maud Marsh, Mary Spink and Bessie Taylor. Sverin Ktosowski alias George Chapman was hanged at Wandsworth Prison on 7th April 1903.

4.8 QUICKSILVER

It is the only metal that is a liquid at room temperature. Many of us are fascinated by the physical properties of mercury, such as its ability to contact a glass surface without wetting it, its capacity for enabling a copper coin to float upon its surface, its amazing mobile nature, and its bright reflective silver appearance.

The latter made it useful for coating onto glass in making mirrors, but due to its toxicity this was discontinued. In addition to those interesting physical properties, mercury also has some chemical properties to fascinate us. Mercury, one of the so-called 'heavy metals', presents itself in many forms. In its metallic state it exists as mercury atoms; the vapour that evaporates from the metal contains mercury molecules; in mercury compounds it exists as positive ions; and in organic mercury compounds the mercury is bonded to an organic molecule.

In each form it shows different levels of toxicity, but we can make some generalisations. Mercury as the liquid is not particularly poisonous, but the vapour it releases is deadly poisonous; the soluble mercury compounds are highly poisonous but the insoluble ones are less toxic. It is the organic mercury compounds that are the worst, as they are treacherously poisonous.

In the environment and in the human body, one mercury compound may react to form a different mercury compound. In this way, a compound with low toxicity may be changed into a compound that is deadly. This is what happened in the Minamata disaster, as we shall see later. Mercury is a sinister poison, for some of its compounds have a delayed effect in the human body. It may be weeks, or even months after exposure to the compound, before the toxic action is felt. It is no exaggeration to say that you don't know you've been poisoned by mercury until you're at death's door.

In nature, mercury is found in all of the above forms. One natural source of the metal is a cave in Spain where metallic mercury drips from the roof and walls. This occurs because the mercury ore in the rock slowly breaks down due to the baking sunshine and releases the metal. Mercury compounds often break down easily when heated, and a traditional extraction method was to roast the mercury ore known as 'cinnabar'. This process has a long history with records showing it pre-dates the Christian era.

Today, alternatives are being found for many applications that traditionally relied upon mercury. It is still used in barometers and, in fact, atmospheric pressure is still quoted in terms of the height of the mercury column, the standard atmospheric pressure being accepted as 760 mm of mercury.

Phasing out mercury is currently underway. For example, alcohol coloured with dye is now being used in glass thermometers instead of mercury. In electrical science, tilt switches still rely upon mercury metal, but the mercury rectifier is a thing of the past. Mercury cells, particularly for small electronic devices and cameras, were used until the late 1990s and did an excellent job, but having mercury cells in electrical toys was not a good idea as a cell could contain up to 45% mercuric oxide.

A notable example of this was the free Spiderman toy that was put into special packs of *Rice Krispies* in America. The product was withdrawn when the technical people at Kellogg's received complaints from the public. A part of the problem with mercury cells is that the cells for domestic use are very small and cannot be sensibly collected for recycling. Many end up in the bin with general refuse and are then dropped into landfill or incinerated. Incineration puts mercury vapour straight into the atmosphere, and landfill allows it to seep out, a process that will continue for many decades to come.

Currently there is the problem of children swallowing the small button cells that are increasingly used in our electronic devices. In general, these pass through the digestive tract and emerge intact at the other end. Sometimes they cause physical problems due to their being a rigid disc that can get sideways and jam. Rarely do they leak their chemical contents, as they are effectively sealed. However, the cheaper ones may be less well made and a seal can fail; deaths are on the increase due to ingested batteries. Fortunately, there are now only a few mercury cells in use.

Mercury is still used in electrolysis of salt solution to make metallic sodium and chlorine, but this also is being phased out. Illegal use of mercury continues in the extraction and purification of gold because of mercury's capacity to form an amalgam, thereby attaching itself to the gold particles, leaving behind the impurity particles. The gold is then purified by separating it from the mercury. The process causes loss of mercury to the environment, hence it being made an illegal operation.

Another amalgam is dental amalgam, which is still used for filling cavities in teeth. It is composed of mercury, copper, silver and tin. It appears surprising that a toxic metal should be used in the mouth where it is in contact with saliva, food and drink. Any mercury leached from the amalgam must pass into the digestive system, but this seems not to be a health risk, although there are concerns.

It is interesting that the main concern over dental amalgam is not when it is in the mouth but when a person with amalgam fillings dies and is cremated. During cremation, the mercury vapour, along with compounds that may form during combustion, is discharged to the atmosphere. Where the corpse is buried the mercury slowly leaches out, along with putrefaction products. The ground water into which the mercury passes may eventually be someone's drinking water.

It is found that mercury metal in the digestive tract is not damaging and simply passes through. However, inhalation of mercury vapour is quite another matter and has poisoned many people in the past. Mercury, if open to the air, releases vapour like any other liquid as it evaporates. If the vapour is inhaled, it is readily absorbed in the respiratory tract.

In this respect, spilt mercury in buildings is a problem. With it being dense and highly mobile it quickly finds its way into cracks between floorboards, where it stays, slowly releasing its poisonous vapour. The rate of evaporation of the mercury depends upon the temperature, and so mercury in the warmth of indoors can build up high concentrations of mercury vapour to be inhaled unknowingly by the occupants.

Mercury is naturally present in the environment and comes from vapour released from minerals in the ground. Human activities also add mercury to the environment, with the main culprit being the burning of coal. Despite the mercury release to the air, the total concentration remains about constant as some of the mercury does not stay in the air but ends up in the soil and water.

A major problem with mercury is that it accumulates in the food chain. Detectable amounts were found in tuna fish some years ago. In order to monitor the situation and pick up hot spots of mercury pollution, the World Health Organisation set the safe level for mercury in the human body at 0.5 $mg\,kg^{-1}$ of body mass. So, for a 70 kg person there could be up to 35 mg of mercury before there is a serious risk to health.

Tuna sales dropped once this became general knowledge, tinted with, or tainted by, media exaggeration. Today tuna offers only a very slight risk. To put things in perspective and look at the situation objectively, we find the risk–benefit analysis is in favour of eating tuna fish because the health benefits outweigh the risks.

It is unfortunate that many mercury applications have been banned. But mercury was on the hit list and had to go. Perhaps there was a degree of over-zealous condemnation of mercury. Unfortunately, for those interested in removing sources of harmful chemicals in the environment, there is nothing they can do about an erupting volcano that is capable of discharging thousands of kg of mercury into the atmosphere.

In the past, compounds of mercury were used in cosmetics, pigments and medicines, but these have also been replaced. For example, Chinese Red was a valued cosmetic pigment and was a popular ingredient for lipstick as it was a bright scarlet-red. The pigment was made by grinding the mineral cinnabar to a fine powder.

The practise of using Chinese Red in cosmetics was stopped a long time ago, but mercury compounds in medicines continued for some time after. Another cosmetic that had a mercury compound as its active ingredient was a skin whitening preparation. The compound was corrosive sublimate, and when applied to the skin it could be absorbed and enter the bloodstream.

Calomel is similar in some ways to corrosive sublimate. Calomel is less poisonous because it is almost insoluble and was once used in teething powders for babies. Giving the baby a dose of mercury to help the teeth break through caused a condition known as 'pink disease' in which the fingers, nose and cheeks took on a pink colouration. One particular powder, Steedman's Teething Powder, contained 26% calomel. Golden eye ointment was, not so long ago, used for treating eye infections and had yellow mercuric oxide as its active component. In the treatment of syphilis, mercury compounds were highly effective at killing the bacterium responsible.

The poisonous nature of mercury compounds makes them useful in agriculture as insecticides and fungicides. In some parts of the World, mercury compounds are popular for treating grain and other seeds destined for planting. In Iraq in the early 1970s wheat seed, intended for planting, was milled into flour and used to make bread, resulting in thousands of mercury poisoning cases of which more than 600 were fatal. The grain had been treated with methyl mercury as a fungicide.

Organic mercury compounds, such as methyl mercury, are extremely poisonous and can be absorbed through the skin with little immediate effect. After some hours, a burning sensation is produced, by which time the mercury compound is attacking the brain tissue. The antidote for mercury poisoning is Dimercaprol, which sequesters the mercury, rendering it inactive and enabling it to be excreted from the body without further damage.

Acute mercury poisoning occurs when soluble mercury compounds come into contact with skin and mucous membranes. These compounds are highly corrosive towards skin, and ingestion leads to severe nausea, vomiting, abdominal pain, bloody faeces and kidney failure. Death occurs within 10 days after exposure. Chronic effects include inflammation of the mouth and gums, excessive salivation, loosening of teeth, kidney damage, muscle tremors, jerky gait, spasms of the extremities, personality changes, depression, irritability, and nervousness. Signs of madness have also been reported, but what human being would not go crazy putting up with such horrendous torture.

4.9 ISAAC NEWTON

Mercury poisoning has certainly been the cause of much illness and many deaths, but there are some people who have survived as their body chemistry has somehow developed tolerance to the poison. For example, Isaac Newton suffered the effects of mercury poisoning. Despite this, he lived to be over 80 and with his well-known mental ability in formulating solutions to scientific problems, it is clear that the mercury did limited damage to his brain.

A case of accidental mercury poisoning occurred in 1810 when a Spanish ship that was carrying mercury sank. The metal was contained in leather bags packed into wooden boxes. HMS Triumph was in the area and went in as part of the salvage operation. The leather bags recovered contained several tonnes of mercury, which were removed from their boxes and stowed on the Triumph. It was not long before some of the bags rotted and released the mercury. With liquid mercury slopping around in the confined and poorly ventilated cargo hold of the ship, there would have been a serious build-up of mercury vapour. The result was that in only three weeks the crew and ship's animals were showing symptoms of mercury poisoning.

4.10 MINAMATA MERCURY

In the 1950s an appalling example of mercury poisoning, from an industrial process, led to the worst pollution disaster anywhere in the World. It occurred at Minamata in Japan where people ate seafood from the local bay into which industrial effluent was discharged. The effluent contained mercuric oxide that had been used as a catalyst in a chemical process, and when discharged it sank to form a sediment. The poisoning was not directly due to the mercuric oxide but resulted from the methyl mercury that was produced by micro-organisms acting upon the sediment.

Methyl mercury is attracted to fat and therefore accumulated in the fatty tissue of the small fish in the bay. When these fish were eaten by the larger predatory fish, the methyl mercury built up in the tissue of the predator. By this means the large fish were concentrating the methyl mercury to many thousands of times the concentration it had been in the water. People consuming the fish caught in the bay were therefore ingesting dangerous concentrations of methyl mercury that went on to build up in the brain and liver with disastrous consequences.

The discharging of toxic waste had gone on from the early 1930s until the late 1960s. It was known to the authorities for many years but no effective action was taken. There are some that might argue that the lack of action meant that this was not an accident. Chronic mercury poisoning resulted and many died. Of those that survived, most were left with permanent damage such as deformity, trembling, tiredness and partial blindness. Several of the women gave birth to brain-damaged children.

4.11 YORKSHIRE WITCH

In terms of deliberate poisoning, mercury seems not to have gained much favour. This is probably due in some part to the strong astringent taste of mercury compounds that would make them immediately noticeable if slipped into someone's food or drink. However, in the early 1800s a Mary Bateman, who was to become known as the Yorkshire Witch (Figure 4.3), killed using mercury compounds. Her motive was to swindle people out of their money and she sought out the vulnerable, particularly those who were seriously in pain and desperate for a cure.

Figure 4.3 Mary Bateman, often referred to as the Yorkshire Witch, poisoned people for financial gain. Her victims were chosen from the vulnerable, particularly those who had severe pain and were desperate for a cure. One of her poisonous potions was corrosive sublimate. She was hanged at York Castle in 1809.
Reproduced with the permission of Special Collections, Leeds University Library/YAS 1607.

One of her potions was corrosive sublimate. Clearly this was not going to make her victim better, but would produce exactly the opposite effect and have the unfortunate person returning for a more effective remedy. Upon the death of one of her victims the authorities were informed. Examination of her victim's body revealed black blisters of gangrene and the stench of decomposition. Bateman's home was searched and a bottle of corrosive sublimate was found along with other poisonous concoctions. Mary Bateman was tried, found guilty and hanged at York Castle in March 1809. The executioner was William Curry who was convicted of sheep stealing and was an inmate in the castle. He remained there to carry out hangings between the years 1802 and 1835.

In 1613, the poet Thomas Overbury was ordered by King James I to be locked up in the Tower of London. Several attempts were made to poison him, including realgar mixed into his broth, giving him an emetic powder containing white arsenic, and getting him to eat some specially baked tarts into which had been secreted corrosive sublimate.

He survived all of these. However, the experiences had left him extremely ill and weakened, which is hardly surprising. The fourth attempt to poison him was with an enema of corrosive sublimate, and this succeeded. The King had the death investigated and those involved brought to trial and the guilty hanged.

In 1898 Roland Miloneux sent mercury poison through the American mail resulting in the death of two of his victims. Mercuric cyanide was his choice of poison and this is somewhat unusual. He was probably thinking this was a double poison and if the cyanide, which should act instantly, failed to kill then the mercury would kick in some hours later and complete the evil deed.

A research chemist, Karen Wetterhol, was accidently poisoned when methyl mercury seeped through her latex gloves during an experiment at Dartmouth College, New Hampshire in 1996. Another accidental mercury poisoning was that of Tony Wimmett who inhaled mercury vapour while in the process of extracting gold from scrapped computer parts in 2008.

4.12 PLUMBISM AND POISON

First smelted around 8000 years ago, lead has found many uses. Whereas gold was warm, bright, scarce, beautiful and associated with nobility and the fine things in life, lead was dull and common but it had great utility, which made it the workhorse of early technology. Gold was ornamental; lead was essential. Not only were the applications known to most people, but the poisonous properties were also recognised. Dioscorides, some 2000 years ago, included lead in his writings on poisons and poisoning.

More than any other metal, lead played a role in the development of civilised living. Its properties make it a versatile material and so it found use for making water pipes, conduits and roofing, drinking vessels, work surfaces, bullets and lead shot. The metal could be alloyed with other metals to make solder for sealing lead coffins, joining lead pipes and for making pewter for tableware. Modern solder used for joining wires and components in electrical circuits is now largely lead free, and risks from inhaling lead fume and dust are minimal. Today's pewter is no longer based on lead and the days have gone when children played with lead soldiers.

Lead was also used as a stylus for writing, which became the lead pencil. Although we still call them 'lead pencils' they are now based upon graphite. This began when, in 1565, large deposits of graphite were found in Cumbria, Britain. The modern pencil is lead free and contains graphite blended with clay to give different hardness ratings. As such the age-old habit of chewing the end of the pencil can now be done without fear of plumbism. It is thought that Beethoven may have suffered lead poisoning, and this theory is consistent with his repeated biting of lead pencils and drinking his favourite Hungarian wine that was sweetened with lead.

4.13 PAINTERS' POISONS

There were often mysterious symptoms associated with artists who worked with oil paint. Artists were prone to melancholy feelings, saturnism and painters' colic. Although the oil itself was relatively harmless, the high concentrations of pigments that produce the colour were not so harmless.

Art historians suggest that Van Gogh suffered from epilepsy and bipolar disorder, and that lead poisoning might have been responsible for his delusions and hallucinations. Michelangelo, it seems, had kidney stones, and paint- and wine-induced gout.

Goya applied paint with his fingers, which is likely cause of his trembling hands, vertigo and tinnitus. In more recent times the Brazilian painter Portinani had a haemorrhage and died at the age of 58 in 1962 after severe bleeding. Doctors had warned him about the risks of lead paints and encouraged him to use the lead-free equivalents, but the artist preferred the lead paints.

Lead was centre stage when it came to pigment choice. Artists in the Renaissance were exposed on a daily basis to toxic compounds of lead. But lead was not the only poisonous metal in the paint pigments. It was often found in combination with other poisonous metals. For example: arsenic; mercury; cadmium; and chromium were popular in pigment chemistry. In considering the illnesses suffered by these painters, we conclude that lead compounds, and those of other metals, were to blame for many a poisoning, but we must guard against blaming lead for every affliction.

Modern awareness of toxic materials has resulted in several of these toxic pigments being replaced by safer alternatives. For example, flake white, which is lead carbonate, has been replaced with titanium white, which is much safer and is based upon titanium dioxide, the modern pigment that now finds use in household paint and as the whitener in paper.

Portraits of Queen Elizabeth I of England who reigned between the years 1558 and 1603 show her face to have a pure white appearance. No doubt some of that whiteness was due to the artist's creativity and his need to err on the side of caution so as not to offend her majesty. As such, the white lead paint of his palette played an important role. Other factors such as the Queen's use of cosmetics are likely. It was common practice to treat the face with a preparation containing white arsenic to improve the complexion. This would be followed by the application of a cream based upon ceruse, which is white lead. This adds up to a lot of poisonous metal going onto the face. It is little wonder that many ladies of those times suffered problems due to poisons.

4.14 RED LEAD IN FOOD

Lead was used in medicines and even as a sweetener for food. Sugar of lead was made in early Roman times by reacting lead metal with sour wine. The sugar we have today is sucrose, but this was not available in early times. Honey was the only available natural sweetener and was too expensive for most people. Sugar of lead was a popular sweetener for wine, and its use as a wine additive persisted up to the Middle Ages.

A recent case of accidental lead poisoning occurred in Hungary in 1994. This was due to red lead being used to boost the colour of paprika powder for putting into sausages and salami. The practice was soon detected and stopped before too much damage was done. It does seem incredible that, in these days of safety awareness, that anyone would think of using lead as a food additive.

4.15 EVIL ETHYL

The increase in the use of lead as civilisations developed, is reflected in the amount of lead found in the environment. In Roman times, smelting of lead ore was carried out in such huge quantities that lead fume and dust from the smelters rose high into the atmosphere and settled out on a global scale. Peat core samples from the moorland areas of Britain show negligible lead in the ground prior to the Roman period when smelting first began. From then on, significant concentrations are found as lead dust fell from the atmosphere.

By studying ice cores going back thousands of years we see lead increasing as more and more lead polluted the atmosphere. The greatest increase in environmental lead was with the onset of industrialisation and, most significantly, when lead began to be added to petrol. The increase in environmental lead meant an increase in human exposure to lead. Increase the amount of lead in the air and there is a parallel increase in the amount of lead in the population, with children being affected the most.

Increases in the human intake of lead were bad news because the metal damages the nervous system. People were being poisoned by exposure, day in day out, to lead in the air. Unlike organic poisons that typically break down and become detoxified in the liver, lead is an element and cannot be broken down but instead accumulates in the bones.

The evidence was available and steps were taken to reduce the amount of lead entering the environment. Unleaded petrol was introduced and now atmospheric lead levels are falling, but there is some way to go before we have an environment that is near to the natural level for lead. In effect, those of us old enough to have lived in the days when cars ran on leaded petrol will have raised levels of lead in our bones, and it's likely some of us suffered chronic lead poisoning. Nowadays the amount of lead in the home environment is minimal, and much of the risk has gone.

However, with the amount of lead and lead compounds that were used in the past we might expect many intentional poisonings, but the literature on poisons shows there have been few. Nearly all the deaths due to lead poisoning have been accidental.

Sugar of lead, with its sweet taste, was chosen by Louisa Jane Taylor in the late 1800s to poison her husband. He was much older than her, in fact he was old enough to be her father, and she had taken to having an affair with a younger man. Shortly after his death, Louisa ran into debt, was evicted from her house and went to stay with a Mr and Mrs Tregillis. The couple were in their 80s and had been friends of Louisa's husband, who they thought had died of natural causes.

Louisa showed affection towards Mrs Tregillis and tended to her needs when she was in bed recovering from having been mugged by a youth. At about the same time, Louisa purchased sugar of lead from a pharmacy. She was familiar with the chemical because some years earlier she had attempted suicide with it. It was probably what sent her husband to his grave. Louisa's purchase of the sugar of lead became a regular occurrence with Louisa claiming that it was to make a solution for a vaginal douche.

Louisa stayed on to look after Mrs Tregillis but her motive was not one of compassion but of monetary gain. The old woman became very ill from time to time but periodically recovered before being knocked down by another bout of illness. On one occasion it was noticed that her face was deathly white, her lips had a strange red colour and her teeth had turned black.

Later, the doctor was called and gave her some medication, but she had more spells of illness, recovery and relapse. Louisa continued to see that she took her medicine. On one occasion Mrs Tregillis complained of the medicine tasting of vinegar and being too sour to take. One medicine she was taking was a fever cure, and this contained nitric acid. For this to taste of vinegar suggests that it contained acetic acid. Acetic acid is formed when sugar of lead reacts with a strong acid such as nitric acid. It would appear that Louisa was mixing sugar of lead with the fever cure. Mrs Tregillis was being poisoned with lead.

The old woman refused further doses of fever cure and Louisa resorted to putting the sugar of lead into her brandy. This way it would not be noticed and so the poisoning continued. On the doctor's next visit, he noted that the woman's gums had a blue line on them that he had come across previously in a case of occupational lead poisoning. He also knew of Louisa buying sugar of lead and became suspicious enough to notify the police surgeon, who confirmed the illness to be due to lead poisoning.

While the case against Louisa was proceeding Mrs Tregillis died. A post mortem revealed lead in her brain and liver, with her stomach containing nearly 30 mg of lead. The trial of Louisa went ahead and the jury returned a verdict of guilty. On 2nd January 1883 Louisa Jane Taylor was hanged for murder. This case shows just how difficult it can be to kill by means of lead poisoning. Mrs Tregillis was in her 80s and was made extremely ill by the lead that she was regularly ingesting, but she amazingly bounced back to near normal health from time to time.

Maybe lead is not quite as poisonous as we think. Perhaps a victim of chronic lead poisoning builds up resistance. It is possible that some people have an inborn tolerance to it. Maybe there are some people whose systems can quickly adapt to excrete the lead? It is a fact that some people do have an incredible resistance to certain poisons. For example, later when considering cyanide, we will learn of the remarkable resistance that Rasputin had towards the poison.

4.16 OLD POPE'S BONES

From time to time forensic analysts are contracted to investigate the remains of people from the past who died under suspicious circumstances. Obviously these are not crime investigations as such, because the aim in investigating crimes is to find the culprits and bring them to justice. What would be the point of establishing the circumstances of a death that occurred hundreds of years ago, because all those that might have been involved in that death are now themselves dead? However, we can learn a lot from the past, and those in historical research sometimes come up with the funds to finance a forensic investigation on human remains from long ago.

Forensic scientists in Bavaria were given permission to take a small sample of bone from the skeleton of Pope Clement II that had been in its stone sarcophagus since just after his death in 1047. The aim was to analyse the bone for indications of lead poisoning. It is known that one way the body excretes heavy metals such as lead is to absorb them into the bone, hair and fingernails. Analysis of many old skeletons has been performed, and the amount of lead present, in non-poisoning case, correlates with the amount of lead in the environment as determined by analysing other materials.

Pope Clement's skeleton contained a concentration of lead that was much greater than normal, possibly indicating that the Pope had died of lead poisoning. Murder is a possibility, but it is also likely that his poisoning resulted from drinking too much wine. In those days, wine was treated with red lead. This is an insoluble compound, but when put into wine it reacts with the acetic acid to form sugar of lead, which is soluble and sweet.

It often happened that wine soured due to the acetic acid being produced and this made it un-saleable. Faced with a significant volume of wine that could not be sold, the temptation must have been to adulterate it with a little extra red lead. Thus, there is no way of knowing whether Clement's death was due to accidental lead poisoning or intentional poisoning. Perhaps with those findings we might enquire as to the amount of wine he was drinking. Lead and its compounds have been used in many different applications and over a long span of time. Perhaps the greatest damage was done when lead tetraethyl was put into petrol.

It would come as no surprise to learn that most of us brought up in the industrial World have some degree of lead poisoning. As a baby I chewed flakes of paint from the skirting board, enjoying the sweetness, which was from lead compounds. I must have suffered some degree of lead poisoning, which is possibly why it has taken me so long to write this book.

4.17 THALLIUM CREAM

This is not well known and is, compared with other metals, a relative newcomer in the history of poisoning. Discovered in 1861 by William Crookes, thallium as the metal finds little use, but its compounds are used in industrial processes especially in the manufacture of lenses, other optical devices and artificial gems. For example, thallium oxide is used for making glass of high refractive index known as 'thallium flint glass'.

Agriculture made use of thallium sulphate as a pesticide for killing cockroaches and rodents. During the early 20th Century thallium acetate creams became popular for removing unwanted hair, and some of them were used in suicides and attempted murder. Thallium acetate was used medicinally for treating ringworm, but the dosage required was dangerously close to the fatal dose.

In the 1930s 14 children died when given thallium acetate for ringworm, because the pharmacy where the medicine was weighed out had faulty weighing scales, resulting in an overdose being supplied.

The symptoms of acute thallium poisoning are: acute nausea; vomiting; diarrhoea; tingling pain in extremities; weakness; coma; convulsions; and death.

Chronic symptoms include pain in the extremities and loss of hair. Among the poisoners who used thallium, the most infamous was Graham Frederick Young. He was serious about poisoning, and the term 'toxicomaniac' was later used to describe him. His poisoning career began early when he was only 12 years old in 1960. He started with tartar emetic and experimented with it on his school friend, who became extremely ill.

Young went on to try another poison, belladonna, using his sister as a guinea pig. She became ill and was taken to hospital where it was found she was suffering from poisoning due to atropine, the active ingredient in belladonna. This caused a lot of trouble for Young but he somehow avoided punishment.

The next to become a victim of his experiments was his stepmother. He began slipping doses of tartar emetic into her food, but then moved on to give her a fatal dose of thallium acetate. The thallium compound was attractive in that it is tasteless, colourless, odourless and it dissolves easily. She died in hospital but the poisoning went by without comment as her death was put down to another cause.

Young then started to poison his father, but the police were on to him. He was apprehended and classified as criminally insane, which sent him to Broadmoor high-security hospital where he stayed until his early release in 1970. During his sentence four suspicious poisonings occurred within the prison. Once free from Broadmoor, he obtained a job as storeman in a Hemel Hempstead company making photographic lenses, a process that involved the use of thallium compounds.

This was just the opportunity Young had been waiting for. Now the serious poisoning could begin, with thallium acetate being used to kill and tartar emetic administered as a means of punishment to those who annoyed him. The authorities eventually caught up with Young who was then sentenced to life imprisonment at Parkhurst Prison where he died at the age of 42. There is proof that he killed three people, but it is believed that he killed 13 victims. In total, the evidence suggests 70 people suffered at the hands of Graham Young, most of them in non-fatal poisonings that caused them to suffer agonising illness.

4.18 SODIUM SALT

We are unlikely to come across metallic sodium in our everyday lives but compounds such as sodium chloride, common salt, are familiar. We put it on our food and we treat the roads with it in winter to prevent icing. It circulates in our bloodstream maintaining a critical level of sodium ions, and it is the basis of the saline drip solutions used in hospitals. In the latter case it is the sodium ions that are important, as they work alongside the potassium ions to ensure the proper functioning of the cell membranes.

There is evidence that excessive salt consumption over a long time causes an increase in blood pressure, which can lead to life-threatening heart problems. This is now recognised as a problem by health authorities and steps are being taken to encourage us to eat less salt. Clearly, salt can be a chronic poison but can it also kill as an acute poison?

Suspecting that this might be the case, we need to look for the evidence. Is it realistic to think that someone could be deliberately poisoned by putting salt into the food? It seems unlikely.

For a large enough quantity to cause death would make the food so salty as to be inedible. Even if the victim were to swallow the food, the normal vomiting mechanism would act and regurgitate it out along with other stomach contents. But where there's a will there's a way.

We learn from the case of Susan Hamilton who put salt solution into a feeding tube that doctors had placed into the stomach of her four-year-old daughter who had a condition that prevented her swallowing properly. It would appear that Hamilton had frequently put salt solution in the feeding tube, but it was in March 2000 that so much salt was administered that the child had a stroke and became permanently brain damaged. Hamilton was found guilty of assault and imprisoned.

4.19 SALT AND CYANIDE

Salt for culinary use is over 99% sodium chloride; the other 1% is made up of trace elements. It is an essential part of our diet and is one of our five senses of taste. The World Health Organisation recommends a daily intake of 5 g sodium chloride, which is equivalent to 2 g of sodium. Most salt is obtained from the underground sediments that occur Worldwide. A smaller amount is extracted from coastal sea water, salt lakes and salt springs. We often distinguish between rock salt and sea salt as if they were quite different substances, but the fact is that both types come from sea water. Rock salt comes from ancient seas that dried up millions of years ago.

Mined rock salt is brought to the surface then ground and sieved. Some of the underground salt deposits use the brine process, in which the rock salt is dissolved while in the ground and the resulting solution is pumped to the surface for purification.

Processing after extraction may involve additives, especially for the purified white table salt. This salt must be made into a powder composed of small crystals so as to be suitable for sprinkling from the salt cellar. To enable good flow properties and prevent clinging due to absorbed moisture, chemicals known as 'anticaking agents' are added.

Two widely used chemicals for this purpose are sodium ferrocyanide and calcium aluminosilicate. The former contains cyanide; the latter contains aluminium. Perhaps this is worrying at first sight, but there is no toxicity issue here. In sodium ferrocyanide the cyanide has a strong chemical bonding to the ferro (the iron). In the aluminosilicate compound the aluminium is tightly bonded to the silicate. As such, neither cyanide ions nor aluminium ions are released, and so these compounds have negligible action in the human digestive system and are readily excreted.

In some of the refined culinary salts, iodide is added to prevent goitre in people suffering hyperthyroidism. The iodide is in minute amounts and provides an iodine level of but a few ppm. All these chemical additives are, in Britain, approved (Committee on Toxicity in Food, Consumer Products and the Environment).

It is interesting to compare the chemicals present in processed salt with sea salt, which has the potential to be contaminated from pollutants in the sea. Taking account of these facts we may wonder which salt is safest. Perhaps the most natural salt is the Himalayan salt mined in Pakistan and India. The source has been sealed deep below ground for the millions of years since it first formed. As such it is untainted by mankind's pernicious pollutants. And the only processes used are crushing and sieving; no chemicals are added. These Himalayan salts also have many trace elements, some of which are desirable micronutrients.

With its pink crystals of different shapes and sizes the Himalayan salt is more interesting than the boring white powder of refined salt. The pink colouration is due to iron. The Himalayan salt and sea salt share a common claim, which is that they are a valuable source of just about every element that the human body needs.

4.20 FLUORIDE

This is highly poisonous and is present in the Earth's crust at 650 ppm. This is an average figure, for in some locations fluoride is found in high concentrations as the compound calcium fluoride in the form of the mineral known as 'fluorite'.

Fluoride from minerals in the ground finds its way into water abstracted as potable water. At 1 ppm in the drinking water, fluoride is effective in reducing dental cavities; at 2 ppm the teeth become mottled; but at 8 ppm a condition known as 'fluorosis' sets in and causes damage to the bones by interfering with the normal calcium chemistry important for bone structure.

In this respect, fluoride behaves as a chronic poison due to small amounts being ingested on a daily basis. However, acute poisoning results where as little as 1 g of sodium fluoride is ingested as a single dose. The first symptoms are nausea, vomiting, diarrhoea and stupor. Swallowing larger amounts is lethal, giving rise to the following sequence: tremors; convulsions; and respiratory and cardiac failure.

It is this highly poisonous property that made sodium fluoride popular as a household insecticide for killing roaches and ants. The deadly poisonous nature of fluoride is often focused upon when the fluoridation of public water supplies is considered as a means of reducing decay in children's teeth. One major worry with this kind of mass medication is when the water is being dosed with sodium fluoride, or other fluoride, and something goes wrong. Or worse still, where someone tampers with the system with malice aforethought.

Mistakes do happen, as we know from the Camelford incident (Britain, 1988), in which people were unknowingly drawing water from their taps that was dangerously contaminated. The problem had started when a delivery of 20 tonnes of aluminium sulphate solution was pumped into the wrong tank at the water treatment works. The chemical was then automatically fed into the purification process. As a result, the water leaving the works, which should have been pure, was most definitely not pure. In fact, it was so impure as to be poisonous, with its aluminium content being 3000 times the safety limit. Fortunately, the problem was soon recognised and there were no fatalities, but there may have been long-term effects.

4.21 EXTRA-TERRESTRIAL TOXINS

It is estimated that 40 000 tonnes of extra-terrestrial dust and debris lands on the Earth's surface each year. This is made up of stardust, comet dust and asteroid dust. In a way, the Earth is a recipient of a load of space waste, but we now get our own back by leaving our litter in space in the form of thousands of bits of junk and disposables from our space technology.

The extra-terrestrial dust has been studied in samples obtained from deep sea sediments and from Antarctic ice. Analysis shows it to be made up of refractory materials such as silicates, along with traces of organic compounds of which some are of biological interest. There are no health concerns over this but if, instead of tiny particles, a large mass of extra-terrestrial matter was to hit the Earth, this would certainly be a health issue. For example, a meteor burning up in Earth's atmosphere or impacting with the ground could wipe out huge numbers of living species.

The Permian Extinction, also known as The Great Dying, is thought to have been due to large meteor impact. It is believed that the collision caused deep-sea methane to erupt, huge amounts of the poisonous gas hydrogen sulphide to be released, and deposits of fossil carbon to burn to carbon dioxide and another poisonous gas, carbon monoxide. With all of this, the global atmosphere would have been seriously poisoned, thus explaining the disappearance of 96% of all life.

In addition to extra-terrestrial materials arriving on Earth, we have high-energy radiation from our sun and other stars. Much of this is absorbed by the atmosphere and so we are protected. However, astronauts and space vehicles get the full blast of it. We have experienced the effect of this when our emissions from CFCs punched a hole in the ozone layer, which is up in the stratosphere.

Ozone absorbs the higher energy ultraviolet rays (UV-B) and protects us. But with the hole in the ozone layer we discovered an increase in radiation-induced skin cancer, cataracts and damaged immune systems. Fortunately, the problems are going away as we no longer use the chlorine-based chemicals as aerosol propellants. The damaged ozone layer is now returning to normal and continues to protect us.

Without the protective effect of our atmosphere, the radiation from the sun would be so intense that life would never have started. Perhaps we should take more care of our precious atmosphere instead of using it as a dumping facility for our waste gases and dusts from our modern lifestyle. Despite the screening effect of the atmosphere, there are some particles originating out in the cosmos and travelling at enormous speeds that get through the atmosphere. They also rip through us, the Earth's crust, its core, out the other side and off back into the cosmos.

4.22 PANSPERMIA

Much of the material that currently arrives on Earth from outer space is from the Keiper belt. And over the millions of years that Earth has existed, there have been impacts from comets and meteorites. The remains of these are largely on the Earth's surface or not far below. Our metals, and the toxic heavy metals that are essential to our industrial societies, are from surface sources, which means they are mainly from extra-terrestrial origins. And the oceans that cover most of our planet probably are a result of water brought here by comets.

The material from space is almost entirely mineral matter such as metal oxides and silicates. But organic matter, containing molecules that may be biologically active, may have arrived with those comets that brought the water. Since the discovery of extremophiles, living organisms that can survive in extreme conditions, there is the possibility that living organisms could have arrived on Earth. It is possible that organisms from fragments of other planets arrived here in the ice of a comet.

And if this is possible, we must consider the chance of other organisms, maybe pathogenic ones, capable of releasing toxins, having arrived on Earth in the past, arriving now or on their way for arrival in the future.

CHAPTER 5

Poison or Medicine

Poisons are medicines; medicines are poisons. It all depends upon the amount that is taken and the length of time it is taken over. Early medicines were substances that had presumably been found by accident to have a beneficial effect such as relieving pain, helping a wound to heel, or providing a sense of relaxation and euphoria suitable for leisure time. It was no doubt a journey of discovery with a trail of dead guinea-pigs along the way.

But, eventually methods of application and correct doses were crudely worked out. The medicines of ancient times would of course have to be drawn from nature's supply of chemicals in plants and animals, as synthetic methods did not make an appearance until modern times. Nowadays, with our knowledge of chemistry and our experience in synthesising new chemicals, we have an enormous range of substance to cure or to simply to relieve the symptoms.

Some might argue that food should join the list of substances because it brings relief from hunger pains, and it prevents starvation. Today we see this quite clearly, for some foods are also capable of creating a high. A modern example is chocolate, which contains many essential nutrients, provides much energy and has chemicals that have a pleasurable psychoactive effect. Wine, also has a similar action but is lower on the nutrition side and higher on the psychoactive scale.

Poisons and Poisonings: Death by Stealth
By Tony Hargreaves

Published by the Royal Society of Chemistry, www.rsc.org

For some ancients, alcohol would have played a part in the everyday diet, for they would likely eat windfall fruit when fresh fruit was not available. And windfall fruit has more often than not fermented, to give a content of alcohol representing the beginnings of wine, albeit somewhat fusty.

For a medicinal drug there is always a specified dosage and a clear warning not to exceed the amount, for the drug then is dangerous – it becomes a poison. With recreational drugs, there is no recommended dosage and many drug users are taking lethal amounts, especially if they are habitual users and have a growing tolerance to the drug. The result is that this group of people have a high mortality rate. With therapeutic drugs, the stated safe dosage can be worryingly near to the overdose level, as is the case for paracetamol.

In our modern World, the classic poisons arsenic, cyanide and strychnine are no longer available for purchase by the general public. However, it is relatively easy to legally obtain therapeutic drugs, even ones normally requiring a prescription. As such, we see that the classic poisons have now been largely replaced by drugs intended for therapeutic use. The public may well have some access to medicines, but this in no way compares with the access that medical people have, whether that is by direct access or *via* a prescription that they can write out themselves. It is no surprise then for us to find among the lists of poisoners a number of doctors and nurses.

5.1 SHIPMAN SERIAL POISONER

Diamorphine is made from morphine, which is extracted from opium, a plant alkaloid, found in the opium poppy, which grows in abundance in hot countries such as Afghanistan. To make the poppy release the morphine, the seed-pods are lacerated resulting in a white liquid known as 'latex' being discharged along the cut. This is the plant's repair mechanism coming into action as the latex dries to seal up the wound. Collecting the opium-rich dried latex is followed by processing that involves extraction using solvents and reaction with acetic anhydride to make the diamorphine.

Morphine and diamorphine are opiate analgesics and represent the strongest known painkillers. Codeine is chemically similar to morphine, but less powerful and used for relief of intermediate pain. Another opiate, papaverine is used to treat erectile impotence, being administered by injection into the penis. Diamorphine and morphine are depressants of the central nervous system and are used medicinally as powerful painkillers. They are administered by injection to people who suffer terrible pain from certain terminal illnesses. Diamorphine is highly addictive due to its power to relax and create a feeling of well-being.

To prevent abuse and dependence, the opiates are legally controlled under the various drug's Acts. Normal street-heroin in Britain is diluted, or 'cut', with white powders such as sugars, caffeine or barbiturates to produce different qualities of street heroin. In general, this illegal source of heroin contains between 30 and 50% diamorphine, but may be as low as 10–15% in the poorest quality material.

Injecting street heroin causes an insidious form of poisoning. It results in addiction from the start; it destroys the personality and the values of a hitherto normal person; it creates a dependence that is expensive to maintain; it pushes its victims into crime. Finally, it kills. If the devil himself were ever to hand out a drug this would be his choice. A typical case of heroin poisoning would be as follows. The body of a young man was found on premises frequented by intravenous drug users. Beside the body there was a hypodermic syringe and some traces of a suspicious brown powder along with empty paper folds of citric acid and discarded needles. The citric acid is harmless but is used to help dissolve the heroin prior to injection.

The pathologist reported that the amount of heroin in the dead man's blood was small and was less likely to be the cause of death than the inhalation of stomach contents. In other words, he died by choking on his own vomit. One effect of heroin is that it induces vomiting. In the forensic analysis of blood in such cases it is morphine that the analysis aims at, since the original diamorphine from the heroin, rapidly breaks down. A small amount of codeine and the metabolite 6-monoacetylmorphine also need to be found to prove heroin use.

Medical use of diamorphine is generally confined to palliative care of those with terminal illness such as cancer, because of its highly addictive nature. However, diamorphine is occasionally used as a muscle relaxant, blood vessel constrictor and local anaesthetic. Side-effects of diamorphine and related opiate drugs include nausea, vomiting, depressed breathing and constipation. Overdose results in coma and death.

Once injected into the bloodstream, diamorphine is quickly converted to morphine, which is then slowly broken down into water-soluble compounds that are excreted in the urine. The presence of diamorphine's breakdown products, in the body and urine, is readily detected by chemical analysis. Diamorphine in bulk powder form may be tested for with spot test kits to assist identification of suspicious substances found at a crime scene or on a suspect.

Diamorphine played a major role in the murders perpetrated by Donald Harvey in America during the years 1983 to 1987. At his trial he claimed to be the 'Angel of Death' as he believed that he was putting his terminally ill patients out of their misery by killing them. However, he later admitted that anger was a motivating factor in some cases. He claimed to have killed 87 but the official figure put the number of killings at between 36 and 57. His chosen poisons were pethidine and morphine although he also tried arsenic, cyanide and insulin.

Beverley Allitt was another Angel of Death but she specialised in the poisoning of babies and young children whilst she was a staff nurse at a paediatric unit of Grantham Hospital. She killed many of her victims by insulin overdose. After some months, her evil activities were discovered and she was charged with murder. At her trial at Nottingham Crown Court in 1993 it was concluded that Allitt's behaviour showed all the signs of a mental disorder known as 'Munchausen's syndrome by proxy' in which harm is inflicted on others to attract medical attention. She was found guilty of murder, attempted murder and grievous bodily harm.

Another case involving Munchausen syndrome by proxy was that of the paediatric nurse Genene Jones in America during the early 1980s. The victim was Chelsea McClellan who had been brought to the hospital for a routine check-up. Jones attended to her and suddenly an emergency developed in which Chelsea stopped breathing and had to be rushed to the emergency room. Chelsea recovered and was taken home but some months later was again brought to the hospital for another routine check-up. Jones attended to Chelsea and gave her an injection. Chelsea stopped breathing and went limp. Jones gave her more injections but soon Chelsea was dead.

A post mortem and investigation was carried out in which the muscle relaxant succinyl choline came to light. Apparently, bottles of this drug had gone missing from the store. It was thought that Jones had killed between 11 and 46 children with injections of succinyl choline and the anticoagulant heparin. In February 1984 Genene Jones was convicted of murder and given a sentence of 99 years.

The case of Dr Harold Frederick Shipman is relatively recent in the history of deliberate poisoning. He had been a general practitioner since the 1970s, he was popular and his patients held him in high regard. In the 1990s there was mounting concern over the large number of deaths among his patients. On close examination of the records it was found that many of the deaths were of elderly women living on their own. Concern turned to suspicion at the sudden death of a fit and healthy woman of 81, Kathleen Grundy. On examination of her will, the whole of her estate was to go to Shipman much to the alarm of her family. The police were called in. Forensic examination of the will focused upon the type print and traced it back to his typewriter. It was proved to be a forgery.

The body of Mrs Grundy was subsequently exhumed and tests carried out. The presence of morphine, from the breakdown of diamorphine, was found. Shipman was arrested and more incriminating evidence came to light. In 1999 he was tried for the murder of 15 of his patients by lethal injection and was given a life sentence. Subsequent enquiries revealed that Shipman's activities as a criminal poisoner had claimed the lives of an estimated 300 patients over a period of 24 years.

Combining opiates with other drugs can produce some especially powerful and lethal concoctions, as was discovered in a case that took place in a Los Angeles hotel in 1982, in which drug dealer Cathy Smith injected the actor, 33-year-old John Belushi, with 11 snowballs. In this context, a 'snowball', also known as a 'speedball', is a mixture of cocaine and heroin. Smith was charged with murder but after some plea bargaining this was reduced to manslaughter. She was sent to a women's prison in California for a short sentence and then was deported to Canada.

5.2 MARILYN'S BARBS

Barbiturates are not found in nature, but are manufactured. The process was discovered by the German chemist Adolf von Baeyer who was working on reactions of urea, a natural chemical present in urine and formed as the final product of protein breakdown. Tests showed that the reaction product (barbituric acid) had mild sedative and anaesthetic properties, which prompted further research and eventually a range of barbiturates became widely available.

Until recently, some popular barbiturates included amobarbital, butobarbital, pentobarbitol and phenobarbital. These have now been largely replaced by safer drugs because of concerns over the addictive nature of barbiturates. This dependence effect became known as 'barbiturism'. However, one barbiturate remains and that is thiopental, which is used as a pre-medication prior to surgery.

The problem with barbiturates is that, if taken on a regular basis, they produce tolerance, carry a serious risk of toxic side-effects, and overdose is fatal. Barbiturates, which are now controlled substances, act to depress the central nervous system and cause intoxication. Symptoms include: confused and slurred speech; sleepiness; and loss of balance and of memory. Withdrawing someone from their barbiturate dependence must be carried out over a few weeks to avoid tremors and convulsions that could be fatal.

When barbiturates are taken in conjunction with other drugs, the danger increases. The well-known example is that of the American actress, singer and model Marilyn Monroe who died on 5th August 1962 at the age of 36 while on a programme of medication using the barbiturate Nembutal, the brand name of pentobarbital, along with chloral hydrate. Chloral hydrate is a powerful sedative and hypnotic drug and can cause dependence.

Like barbiturates, the chloral can also lead to dependence. It is normally given by mouth as a syrup, particularly for children and the elderly. Primarily used to induce sleep, chloral hydrate is not without its side-effects, which include nausea, vomiting and gastro-intestinal problems. It is also a reactive drug and can undergo adverse chemical reactions with other drugs that are in the system at the same time.

Although the official cause of death was 'suicide due to barbiturate overdose' there are many who are not convinced and different theories are alive. Each theory arrives at a different conclusion: suicide; murder; or medical accident. One theory claimed that Marilyn was involved with the Kennedys, President J. F. Kennedy and his brother John, creating an embarrassment for the American government.

On reading through the accounts written about her death there certainly appear to be many things that would these days, and possibly then, raise a few eyebrows if relied upon in court. For example, her phone-call records had been wiped, certain times relating to the hours around her death were changed and, soon after her death, a key witness made a rapid departure to Europe.

An outline of the established facts follows. Sergeant Jack Clemmons of the Los Angeles Police Department received the phone call at 04:35 am that reported Marilyn Monroe had committed suicide. Clemmons went immediately to Marilyn's address where, in the bedroom, he saw her body lying nude and face down. In the room was Ralph Greenson her psychiatrist, and Hyman Engelberg her physician.

The live-in housekeeper, Eunice Murray was also in the house and she had the washing machine on and was busy doing some laundry. Clemmons noted this as it was perhaps a little unusual to be doing the washing at such an early hour. Clemmons received confused accounts of the events of that night. In particular Eunice was evasive and changed her mind about when she discovered Marilyn's body.

By 08:00 am Marilyn's body was at the city morgue, and soon after Dr Thomas Noguchi was performing the post mortem. He was accompanied by John Minor who was an expert in assessing suicides. The body was examined thoroughly for external injuries or needle injection marks of which none were found. Forensic analysis gave some interesting results. In the blood there was 8 mg of chloral hydrate and 4.5 mg of Nembutal, but the liver contained Nembutal at a level of 13 mg per 100 mL. The ratio of Nembutal in the blood to that in the liver suggested that Marilyn must have been alive for several hours after the ingestion of Nembutal, as most of it had been removed from the bloodstream and was being broken down in the liver. Other tests showed no trace of Nembutal in her stomach or duodenum.

A major area of Marilyn's colon had an abnormal discolouration that was consistent with rectal administration of barbiturates or chloral hydrate. Greenson had stopped prescribing Nembutal and switched to chloral to help her sleep. Unfortunately, Engelberg gave Marilyn a prescription for 25 Nembutal tablets and these were in addition to a store of chloral in gelatine capsules, without Greenson knowing. The exact number of pills at Marilyn's home during the last phases of her life is confused and partly a result of failure to communicate on the part of Greenson and Engelberg.

It is known that when Marilyn was awake, she was difficult to manage and was resisting oral medication. Greenson, it would appear, decided on administration of a heavy dose of chloral by means of enema but he delegated the task to Eunice, a woman who had no medical training. Administering an enema to Marilyn would not have been a problem as she was used to enemas as a part of her need to remain slim. A dangerous situation had been created in which the chloral hydrate would adversely react with the Nembutal. The administration of an enema can be a messy affair. Was this why Eunice was doing some laundering at such an early hour?

The circumstances surrounding this death were somewhat curious and certainly more involved than many other deaths by poisoning. Had Marilyn not been a top celebrity, the verdict of suicide would have rested. However, whenever a celebrity suffers an untimely death there are always those who take it upon themselves to challenge the cause and manner of death, or challenge the legal processes and evidence that led to that verdict.

There may be genuine reasons for this but often it is an action motivated by the desire for involvement: a need for a claim-to-fame, rather than a need to see justice done. In this respect we are reminded of the other deaths attracting conspiracy theorists: Napoleon's arsenic poisoning; Princess Diana's fatal car crash; and Michael Jackson's death. The conspiracy theorists are always ready to work. And by the time you read this there will be others waiting to climb on the bandwagon.

Conspiracy theories aside, there is one thing that the Monroe case demonstrates. That is, how difficult it is to arrive at a conclusion when there is so much evidence to consider. It is, if we look at it from a mathematician's viewpoint, a complicated equation with many different variables, and with different levels of confidence about the integrity of each one. Demonstrating that a particular sequence of events took place by processing a multiplicity of evidence cannot be an easy task.

A few years before the death of Marilyn Monroe, barbiturates were used along with morphine by Dr John Bodkin Adams. He killed Gertrude Hullett in 1956 at Eastbourne. Apparently she was depressed and told Adams she wanted to commit suicide. Adams was suspected of killing a total of 163 patients with morphine and barbiturates.

Another barbiturate poisoning was that of Brian Epstein, aged 32. He was the manager of the group called the Beatles and died of Carbitral poisoning in 1967. He was taking the drug on a regular basis in the form of a sleeping pill, but he was in the habit of using large doses, and had likely developed a tolerance for it. On that occasion he had taken the pills in combination with alcohol. Alcohol reduces the body's tolerance to barbiturates and in Epstein's case it was reduced to a fatal level. Carbitral is a central nervous system depressant, as is alcohol.

Carbitral is classed as a sedative–hypnotic. After Epstein's death this barbiturate was banned. At the inquest it was recorded as an 'accidental death'. Judy Garland also died after taking a large dose of barbiturates and drinking alcohol. Other famous people who died of barbiturate overdose are: Jean Seberg; Jimi Hendrix; and Kenneth Williams.

5.3 TRUTH DRUG AND SEVERED HEAD

Hyoscine is also known as 'scopolamine' and has the trade names Buscopan and Scopoderm. It is sometimes referred to as the 'truth drug'. It is a plant alkaloid and related to strychnine, nicotine, aconitine, morphine and others. Crude hyoscine is extracted from the plant henbane as a viscous water-soluble liquid that is unstable and easily breaks down on standing, or in conditions that are acidic or basic.

With its powerful effect upon the nervous system it can block essential functions. It behaves as a hallucinogen at certain dosages and it was once popular to relax the uterus during labour, for treating motion sickness, and preventing period pains. In mental institutions such as Bedlam it was given by mouth or by injection to quieten the violently insane. Symptoms of an overdose are throat dryness, delirium, stupor, coma, paralysis and death. Where a person survives an overdose, they make a full recovery and apparently have no recollection of what had happened to them.

Dr Hawley Harvey Crippen was a homeopathic doctor and had a range of remedies to hand. He purchased five grains of hyoscine as hyoscine hydrobromide. It seemed that he wanted to sedate his wife while he entertained Ethel, his typist, in a separate room. Although he was unhappily married to his wife Belle, it seemed it was not his intention to kill her. However, he accidentally gave her an overdose from which she became delirious and rampaged around the house shouting.

In panic he shot her in the head and, in his confused reasoning, decided to remove the head. The remainder of the body he cut into small pieces and buried beneath the cellar floor. But Belle's head, presumably with a bullet lodged inside it, was never found. Crippen and Ethel then made a hasty departure, setting sail for America, but the police caught up with them. Crippen was tried, found guilty and hanged at Pentonville Prison in 1910 but Ethel was acquitted.

5.4 LETHAL BUT LEGAL

The element potassium is to be found in a variety of compounds, particularly the salts potassium nitrate and potassium carbonate. In these compounds the potassium exists as ions that are released when the salts dissolve in water or in body fluids. Potassium ions, along with sodium ions play a vital role in maintaining the proper function of the body's cells. The cells work perfectly well when the level of potassium in the bloodstream supplying the cells has a concentration that is low.

Changing this concentration results in the cell chemistry being disrupted. In the case of nerve cells and muscle cells, the effect can be dangerous and even fatal. In cases of kidney failure, which may itself have resulted from poisoning, the blood fails to be properly filtered and a high potassium level occurs. This, in turn, causes the heart rhythm to fail, leading to cardiac arrest.

When the potassium concentration falls from the safe level due to fluid loss through diarrhoea or vomiting, muscle paralysis occurs. Clearly the body is highly sensitive to potassium in the blood. It is however, less sensitive to ingested potassium. For example, many people these days, in order to lower their sodium intake for health reasons, use low-sodium table salt. This is a mixture of sodium chloride and potassium chloride, and provides the same level of saltiness as normal table salt that is entirely sodium chloride. And potassium chloride mixed with sodium chloride is the basis of the rehydration treatment that we take orally after a serious bout of diarrhoea.

Thus, potassium is fairly innocuous when ingested, but it is lethal if injected into the bloodstream unless under strict medical supervision as part of treatment for potassium deficiency. Although swallowing potassium ions is relatively safe when those ions come from, say, potassium chloride, an entirely different situation exists if the potassium ions come from potassium cyanide, for the cyanide itself is deadly poisonous.

Injection of potassium ions into the bloodstream is used in executing murderers in America and China. The procedure does not simply involve injecting with potassium, but has three stages. In the first stage, a saline solution is dripped into the bloodstream by means of a tube connected to one arm. This is followed by pentobarbital, a barbiturate anaesthetic, which is also administered through the tube. In normal surgery, the anaesthetic level of pentobarbital administered intravenously is around 0.1 g. For execution the amount is 5 g, ensuring that the condemned man feels nothing. This huge dose would be sufficient to knock out a person for a couple of days. In less than a minute the condemned man is unconscious, and in some cases he has already died from the overdose of barbiturate.

To make sure death has occurred, there is a further step in which pancuronium bromide, with trade name Pavulon, is fed into the tube. This is a paralysing agent and works by preventing the diaphragm and other respiratory muscles from working. Asphyxiation follows and breathing ceases. In surgery, Pavulon is used in minute amounts as a muscle relaxant, but in execution the dose is massive. Finally, potassium chloride is given *via* the tube and the high level of potassium ions guarantees total cardiac arrest. Death occurs from 5 to 18 minutes from the start of the execution. Euthanasia has similarities with the above procedure. The patient normally takes the Pentothal (thiopental) anaesthetic orally but, for some illnesses this is unsuitable and so intravenous injection is used. When the patient is unconscious, a Pavulon injection follows.

Knowing full well the effects of potassium injection, Dr Michael Sevango administered lethal doses of potassium chloride solution to patients. He pleaded guilty to deliberately causing four deaths by injection of potassium, but it was estimated he killed 35 people.

5.5 PENNYROYAL

Euthanasia kills people who are deemed to have such a poor quality of life that they are better off dead. This usually applies to the old, but at the other end of the scale we have abortion. Significant in both situations is the decision by others to plan intentional killing, with poison chosen as the favoured weapon.

Preventing unplanned pregnancy has been a major concern since early times.

The evidence for this comes in the form many weird and wonderful descriptions of methods for preventing fertilisation. Approaches to dealing with the situation have included contraception and, failing that, abortion or even suicide. In each instance poisons have been shown to play a major role. Contraception often involves substances for killing the sperm and known as 'spermicides'. Pessaries have a long history of being a convenient means of introducing a substance to destroy sperm cells. One early pessary was made from crocodile dung and honey and was described in Eber's Papyrus as far back as 1550 BC.

However, the killing of sperm is another story. Here we are concerned with the killing of humans. Abortion, if performed before a certain stage, is now legal in many countries. There are physical means for destroying the foetus and there is administration of certain substances that will poison the system upon which the foetus relies, thereby killing it.

Chemicals that perform this function are known as 'abortifacients' and, in addition to contraceptives, have a lengthy history. It was not so long ago that the local 'wise woman' was the source of things to cure anything, as she had on offer a potion for every occasion. Herbs were her main raw materials and one in particular was successful in preventing the development of the fertilized egg, if taken at an early enough stage. The herb was the well-known pennyroyal, a type of mint that is readily available. In fact, it is still used today, unofficially, for inducing a period that is overdue.

Mashing a few leaves of the herb with hot water produces a tea that contains low concentrations of the compounds pulegone and menthone. If the brew is too strong or swallowed in large doses, it can be fatal. The proper medical termination of pregnancy is today performed by oral administration of mifepristone, which produces an abortion if taken in the first 20 weeks of pregnancy. The drug works by blocking the action of progesterone, the hormone essential for maintaining pregnancy. This is followed, after 36 hours, by gemeprost, administered intravaginally to induce uterine contraction.

The local wise women could also be relied upon for love potions, spells and incantations. She was clearly into the magical arts and that must have re-inforced the idea that herbal concoctions carried some mysterious force, an evil power drawn from the earth and emanating from the devil. This was especially so if she used fungi, because these plants were thought to be the devil's creation since they appeared at night time from the damp black earth and were loaded with deadly poisons. It is little wonder then, that occasionally a wise woman was held to be a witch, and summarily burnt at the stake or subjected to some other inhumane punishment. Notions that the wise woman was in league with the devil would be re-inforced by the spells and incantations that were also a part of her stock in trade.

Not only did her patients run the risk of physical poisoning from her concoctions, but they were also at risk of having their minds poisoned. Minds are delicate things and can easily be knocked off balance. Imagine someone in say, Medieval times, believing that a curse had been put upon them. Today we might consign this to the bin marked 'bunkum'. What a load on nonsense. Our judgements suit our belief system. But is it complete nonsense? Not so long ago, this was real. It was real in the mind of the believer, just as real as God is in the mind of the religious. Belief is still alive and well, and controlling what we do. We see this in drug trials, where a placebo with no medically active ingredient produces positive results. The patient has faith in its powers and so there is a positive outcome.

5.6 NAME YOUR POISON

"Name your poison". It's an old saying dating back to the early 18th Century. Even then, before the days of safe-drinking limits, alcohol was recognised as a poison, which could be acute, leading quickly to death or which could have a chronic effect such as progressive liver damage to the point of failure.

We might consider whether or not alcohol is a natural part of our diet. In thinking of alcohol in some depth, we need to be precise about its chemistry. The alcohol in alcoholic beverages is ethanol. The word 'alcohol' is a generic term for a group of related compounds. For example: methanol; ethanol; propanol; butanol; and others. The other alcohols are not for drinking. In fact, methanol is extremely poisonous as we shall see later. Ethanol as part of the human diet goes back to early man, who lived a nomadic life and survived as a hunter gatherer. The gathering would have included fungi, berries, nuts and fruit. Not all the food gathered would have been fresh, from which it is reasonable to conclude that windfall fruit would have played a part in his diet.

This fruit would be in various states of fermentation and decay. It would have been recognised that at a certain point in the fruit's decay process, the taste was bearable. Beyond the taste it would have had an intoxicating effect when ingested. Perhaps it became a favourite, a treat, something to seek out. We might suggest that the seeds of our alcoholism go back primitive man.

But it is not only humans that enjoy the effect of ethanol. For example, cows being fed their winter diet of silage, seem to be rather more content and relaxed than they are when eating fresh grass. And it is not unusual for them to be a little unsteady on their feet and stagger a little. The explanation is that they have become intoxicated by eating the silage, which is hardly surprising as it is fermented grass. Some of the sugars in the grass are turned into ethanol.

It would seem that in moderation, ethanol is desirable. But moderation is hard to define. Let us consider the evidence. In Britain around 500 people each week arrive at, or are carried, to Accident and Emergency departments of hospitals suffering ethanol poisoning. Today we recognise ethanol as being top of the list of poisons. Chronic ethanol poisoning results in a huge number of premature deaths Worldwide. Acute poisoning by ethanol sends many a binge drinker to an early grave. Many countries attempt regulation of alcoholic beverages, often for religious reasons, and this is nothing new. Records show that in early China, around 2000 BC, prohibition was introduced.

A common approach is to have a total ban on these drinks, but this has its problems as was seen when prohibition was introduced in America in the years between 1920 and 1933. It is curious that this prohibition did not make it illegal to consume alcoholic beverages, but outlawed the manufacture and distribution of them.

Those were the days when organised crime took control of the booze industry and with horrific results. A major problem with the illicit drink was that methanol, the most toxic of all the alcohols, was being added. Methanol, or 'wood alcohol', is a cheap industrial solvent. It can be blended with drinks containing ethanol and is not easily detected by its taste. The methanol has an intoxicating effect like that of ethanol, but it is highly poisonous.

The consumption of cheap alcoholic beverages containing methanol is a growing problem today. It is especially bad in the poorer communities, as can be seen in the following examples: in June 2015, 90 people died in Mumbai; in December 2011, 126 people died in West Bengal. Adding methanol to beverages is illegal in nearly all countries concerned, but it still goes on. It is understandable that these people take risks when we consider the cost of drink. In India, a 700 mL bottle of whisky costs £4.80; a bottle of the illegal stuff, known as 'hooch', is a fraction of the price.

It is interesting to note that the solvent sold in Britain as 'methylated spirit', and commonly called 'meths', is ethanol that has been intentionally poisoned. Methylated spirit is mostly ethanol with a few nasty chemicals added to it. The dye methyl violet gives it a purple colour, pyridine provides a repulsive smell and taste, and methanol, which gives it the name 'methylated', makes it poisonous.

The idea is to enable it to be sold as a cheap household solvent or as fuel for spirit burners, but to discourage people from drinking it. It is cheap because, compared with ethanol for drinking, it carries no alcohol duty. If it was sold in its unadulterated state, many would drink it rather than the real stuff and the Chancellor of the Exchequer would be deprived of his dues. Despite its poisonous nature there are the intrepid few who do drink meths. Some tolerance may develop, but generally the drinker is poisoned by the additives. The ethanol inebriates, the pyridine causes sexual sterility, and the methanol turns into the poisonous acid, formic acid.

Ethanol is often used in preparing medications because it ensures the active ingredients remain in solution. If the medication is to be taken by mouth, the alcohol is ethanol, whereas propanol is used in preparations that are to be applied to the skin. It is noteworthy that Gripe Water, which is given to babies to relieve colic, contains ethanol. The original formulation had 3.6% ethanol, which is about the strength of regular beer. But now the ethanol content has been reduced, and the more modern formulations contain no ethanol.

Ethanol plays a big part in the lives of people in many countries. In fact, it is so entrenched in many cultures that it is not thought of as a recreational drug. To suggest to a regular drinker that he is a drug addict dependent upon a psychoactive drug, would attract an immediate denial and even some indignant response. The reasoning seems to be that alcohol is legal; the drinking of it is an adult activity and it is promoted by advertising. And governments sanction it by collecting tax on it. On the other hand, drugs are illegal, expensive and are generally for the younger ones, especially teenagers.

5.7 HOOPER'S HOOCH

Some years ago the drinks industry introduced a range of intoxicating drinks that would appeal to the younger drinker. Research had shown that the bitter taste and the sting of ethanol put off some youngsters. There was much business to be had by marketing fruity and brightly coloured drinks with flavour at high enough levels to disguise the ethanol. In the 1990s alcopops came onto the market with ethanol contents of 4 to 7.5%, and with some stronger ones at 12.5%. Hooper's Hooch and Two Dogs soon became firm favourites with under-age drinkers.

As we see every day, ethanol and other recreational drugs have a major impact upon health and crime. It is an unfortunate aspect of modern living that as disposable income increases, we consume more alcohol, swallow more psychoactive pills, and devour bigger heaps of food than ever before. And the alcoholism, drug addiction and obesity are there as evidence. In this respect, we see that as standards of living increase, the quality of life decreases. A kind of inverse relationship exists. However, we must be cautious about conclusions for a relationship does not necessarily indicate 'cause and effect'; so we must reserve judgement.

Alcohol is a major factor in crime, violence and motor accidents. In fact, it was the latter that resulted in the Road Safety Act of 1967 and the introduction of roadside screening to test drivers for alcohol, in the hope of reducing the carnage from drink-drive accidents. If ethanol was something new and under consideration for sale to the public, it would most certainly be prohibited. Ethanol causes intoxication, that is, it has a toxic effect on the body. We might be tempted to think of ethanol as a stimulant, but the fact is that ethanol is a depressant. To understand more on the toxic effects of ethanol we need to consider the figures. Different drinks contain different concentrations of ethanol.

The concentration is usually quoted as 'percentage volume' or, more recently, as the number of 'units' where one unit (Britain) is equivalent to 10 mL of pure ethanol. The strength of alcoholic beverages is given as 'alcohol by volume; (%ABV) which means the number of mL of pure ethanol in 100 mL of beverage. Typical ethanol contents of some popular beverages are: regular bitter 4%; strong lager 5%; strong cider 7.5%; red wine 12.5%; sherry 15%; and whisky 40%.

The stated maximum introduced in January 2016 is 14 units for both men and women. Within these limits, ethanol is regarded as beneficial. It benefits the drinker and it benefits the excise man. Is this safe? Is it desirable to occasionally stress the body by exposure to toxins? Here we might reflect upon the habit of King Mithridates as he dosed himself with poisons to be build up resistance. There may be parallels with the modern problem of allergies, and the tendency of more and more children to be vulnerable. In our enthusiasm to make the World a perfectly place safe to be, are we not in danger of letting the body's natural ability to adjust to new challenges become redundant?

5.8 ETHANOLIC EXCRETIONS

Studying the factors that influence the absorption of ethanol enables an understanding of how alcohol behaves in the body, and the poisoning action it produces. When we look at the ingestion of ethanol, we note some important factors that have a bearing upon the state of intoxication. Obviously, the strength, the volume, and the rate at which it is drunk are major factors.

But other conditions play a part. For example, drinking on an empty stomach enables a faster rate of alcohol absorption than drinking on a full stomach. Drinking carbonated drinks provides for quicker passage of alcohol through the lining of the small intestine, than occurs with non-carbonated alcoholic beverages. This is one reason for the whisky drinker taking it with a splash of soda water. It's the fizz that makes the difference, because the presence of carbon dioxide speeds the up the absorption of ethanol into the bloodstream. Body mass, gender and state of health also have an influence.

The body recognises ethanol as a poisonous substance and uses various mechanisms to eliminate that poison. Thus, elimination is as follows: 95% is broken down in the liver by the enzyme ethanol dehydrogenase; 5% is lost in exhalation and urination; and a minute amount is given out in perspiration. A male of high body mass will eliminate the ethanol at over 20 mL h^{-1}. For a lower body mass or for an inexperienced drinker the rate is less. In general, the elimination rate is about 10 mL h^{-1}. It also depends upon how much enzyme the drinker's body is capable of producing. In certain ethnic groups that do not traditionally drink alcoholic beverages, there is a lower capacity for producing this enzyme. The drinking of even a small amount of alcohol can quickly intoxicate them.

An understanding of how quickly people sober up is important in the conviction of drunken drivers, and the police can perform a back-calculation to extrapolate a driver's blood alcohol concentration to the time an incident occurred. The formula is in the Glossary.

5.9 FOOD OR FUEL

The ethanol for alcoholic beverages comes from fermentation of sugars from grapes or grain. In the production of spirits, a fermented solution is taken and subjected to distillation. For example, brandy is made by distilling red wine, and whisky is a product of distillation of fermented grain.

Ethanol is a good fuel. It will burn in a spirit burner producing plenty of heat and it provides the human body with lots of calories. Comparing pure ethanol with other food substances we have (kcal g^{-1}): ethanol = 7.0; fat = 9.0; carbohydrate = 4.0; and protein = 4.0.

Ethanol is such a good fuel that it is now being used as a petrol replacement. In South America a product known as Gasohol is now widely used. The selling point is that the ethanol used is made by fermentation of plant sugars. As the plants grow they remove carbon dioxide from the atmosphere by photosynthesis to build up the starch, sugars and cellulose. When the sugars are fermented they make ethanol which, after purifying by distillation, is suitable as a fuel. Of course when that fuel burns it releases carbon dioxide into the air. In effect, when we view the whole process there is no net change in atmospheric carbon dioxide. It is a sustainable process and a steady-state exists. This contrasts with burning fossil fuels, which put carbon into the air that had hitherto been locked away underground.

But you never get something for nothing. Ethanol, like most other bio-fuels, takes as its starting material plant materials, and therefore is in competition with the demands of food production. Producing ethanol from plant sugars is destined to increase not only for beverages but also for making chemicals, which are currently made from fossil carbon. It seems that ethanol is the one poison that will play an increasingly important role in our future. We might be left to wonder just how much of this ethanol will be diverted to illicit beverages.

5.10 TYLENOL TAMPERING

Paracetamol, known in America as Tylenol, is an analgesic and antipyretic, and comes in pill form for general use. The active ingredient is acetaminophen which was introduced in the 1950s, when it was discovered how to make it from phenol. Often the drug is formulated with other analgesics for non-prescription sales. For example, Anadin Extra contains aspirin, paracetamol and caffeine.

An unfortunate aspect of paracetamol is that the therapeutic dose is quite near to the lethal dose, making it easy to overdose by accident. Paracetamol taken in high dosage causes liver failure and death after about three days. If the victim of paracetamol poisoning can be taken to hospital, tests may be performed to measure the amount of paracetamol in the blood, and a decision made as to whether treatment with acetyl cysteine is appropriate. This drug can protect the liver from the damage that the paracetamol inflicts. Paracetamol, taken in combination with an alcoholic drink, is now a common means of suicide. What is not appreciated by the person attempting suicide is that it does not kill quickly. There may be two or three days of shear agony before the final moment arrives.

In Chicago in 1982 paracetamol sold as Extra Strength Tylenol killed seven people, but not through overdose. The fatalities occurred due to cyanide poisoning. Investigations revealed that the product had been tampered with by the addition of potassium cyanide, after leaving the factory. The culprit, clearly intent upon killing innocent people at random, was never caught. Following this poisoning, a series of copy-cat crimes began, and this forced the introduction of antitamper packaging.

CHAPTER 6

Man-made Menace

There is a tendency to think of chemicals as substances that are highly reactive and, because of this property, are damaging to living systems and the environment. Despite our negative thoughts we grudgingly accept that chemicals are essential for industry to manufacture the materials needed for modern living.

For example, we would all recognise sulphuric acid as being highly corrosive, and capable of burning its way through many materials including human flesh. The evil deeds of John George Haigh come to mind here. He disposed of the bodies of his murder victims by dissolving them in sulphuric acid. This became known as the 'acid bath murders'. Certainly this acid is treacherous stuff, but we know it is essential. Car batteries need it, detergents are made from it and it is the starting material for other chemicals.

But would we regard water as a chemical? Probably not. Water is not seen as something highly reactive and capable of destroying living systems or poisoning the environment. In fact, the human body is itself mainly water, as is much of our environment. The fact is that water has claimed more lives than all those industrial chemicals put together. In death by drowning a person inhales water, which stops the lungs working. In turn, essential cells are starved of oxygen and the poisonous chemicals from their respiration build up and kill them.

Poisons and Poisonings: Death by Stealth
By Tony Hargreaves

Published by the Royal Society of Chemistry, www.rsc.org

Our modern World relies not only upon chemicals from industry, but also on natural chemicals. There are man-made chemicals and nature-made chemicals, as well as some that are a bit of both. The World's favourite drug, alcohol, may be made in a chemical works from ethene (from petroleum) or it may be made by nature from plant sugars in the process we call 'fermentation'. The ethanol from the chemical works is chemically the same as that from fermentation. It is fermentation that gives us our alcoholic drinks. Whether man-made or nature-made there are chemicals that are reactive and dangerous alongside those that are benign.

Manufactured chemicals are needed for a range of applications. Some of this manufacturing activity is aimed at making poisons for killing living systems such as insects, larvae, rodents, molluscs, fungi and weeds. Regrettably, the chemical plants that produce chemicals for constructive purposes can often switch their production to chemicals that are destructive. Making cyanide to kill rats is no different to making cyanide to kill humans. The concentration camps of World War II Germany are a horrific example of this. There are also many countries that can switch their chemical manufacturing plants to turn out chemical warfare poisons for mass killings of humans. The production of pesticides has close connections with making weapons of mass destruction.

With a large chemical industry, it is not surprising that accidents occur and dangerous chemicals leak into the environment and cause poisoning. The Union Carbide works in Bhopal, India released methyl isocyanate into the atmosphere in 1984. It caused around 8000 deaths and left approximately 20 000 people with serious health problems related to isocyanate poisoning. The plant, which was manufacturing carbaryl pesticide was claimed to have failed due to defective maintenance, although sabotage was also likely.

In Japan, mercury from a chemical works was discharged into Minamata Bay allowing the heavy metal to enter the food chain. Six tonnes of toxic gas containing dioxins was released when a reaction vessel ran out of control at a chemical works in Seveso, Italy in 1976. The poison settled out over a wide area. People were warned not eat the locally grown fruit and vegetables, and thousands of animals were slaughtered to prevent the chemicals entering the food chain.

Nature also spills out poisonous chemicals and damages the environment. When a volcano erupts, it discharges huge quantities of poisonous heavy metals and massive amounts of sulphur dioxide, which reacts with oxygen and rainwater to form sulphuric acid – nature's own acid rain. Among the metals emitted is the highly toxic mercury. It is a worrying thought that there are many poisons being manufactured and moved around the World in large quantities. From time to time, some of these poisons, perhaps only small amounts, fall into the wrong hands or mysteriously disappear.

6.1 CYANIDE

In the past, cyanide in the form of sodium cyanide or potassium cyanide was generally available and widely used as a pesticide. Rodent poisons were often based upon these cyanides. These chemicals are no longer available to the general public, but are important in industry. They are used in electroplating of metals and the case-hardening of steel, during which the metal part is immersed into molten sodium cyanide.

Gold and silver are extracted from ores using cyanides, and fruit trees are fumigated with cyanides. Both the sodium and potassium salts are highly soluble, which means, if ingested, they rapidly release cyanide ions into the bloodstream. Cyanides can be absorbed by the body through injured skin and through inhalation of dust from handling the cyanides, or from breathing prussic acid gas released from these toxic chemicals.

Cyanide is highly toxic when swallowed, leading to vomiting, convulsions, coma and death. The whole of the process takes a matter of minutes. Evidence of cyanide poisoning is the smell of almonds on the breath of the victim. The antidote for cyanide poisoning is treatment with dicobalt edetate or a combination of sodium nitrite and sodium thiosulphate. This antidote has saved many lives, but it must be given to the patient immediately. Cyanide poisoning is indicated by the appearance of a deep-red colouration to the skin. This is not to be confused with carbon monoxide poisoning, which produces a cherry-pink colour.

A simple test for cyanide is to treat tissue samples with silver nitrate. The appearance of a white precipitate indicates the possibility of cyanide, but it is not proof and further tests would be needed. For example, if the precipitate is reacted with ammonium sulphide and ferric chloride, it produces a scarlet solution that suggests a high probability of cyanide.

Prussic acid (hydrogen cyanide) is a gas released when cyanide salts come into contact with acids. The gas, being extremely toxic, poisons when only a tiny amount is inhaled. Cyanide salts such as those of potassium and sodium are poisonous, because they release cyanide ions when in contact with water. Prussic acid gas also releases cyanide ions, but as the gas is inhaled these are rapidly absorbed by the lungs and go straight into the bloodstream.

Many everyday plants contain cyanide, but combined as the less toxic cyanogenic glycoside, rather than as cyanide ions. The kernels of bitter almonds and cherries are especially rich in this form of cyanide. Leaves, such as those from the laurel bush contain cyanide. It is thought that Graham Young made extracts of such leaves, which he had collected from a laurel bush in the Broadmoor high-security hospital where he was being held. It is thought he tried out his extracts on fellow inmates.

Cyanogenic glycosides, like those in the laurel leaves, have the potential to be converted into cyanide ions under the right conditions. The human body can deal with small amounts of cyanogenic glycoside in food by breaking it down into harmless substances that are then excreted. However, 1 kg of marzipan is enough to kill a man. But this would have to be eaten in one go. If ingested it would be soon removed by vomiting. As a means of poisoning someone, offering them a large slab of marzipan is unlikely to succeed as a weapon of stealth.

6.2 PRUSSIC POISON

Inhalation of prussic acid gas proves fatal in less than a minute. The deadly poisonous nature of prussic acid may have killed the Swedish chemist, Scheele when he inhaled a small amount of it while he was preparing some as a sample.

Prussic acid was used in the gas chambers of Auschwitz and Birkenau in Nazi Germany. It was made when the industrial chemical Zyklon B was reacted with acid. The Zyklon B was pellets of sodium cyanide and was manufactured as a pesticide. It was used to kill 1.2 million people. At the fall of Berlin, Adolf Hitler and Eva Braun committed suicide by swallowing cyanide pills.

There have been cases of potassium cyanide or sodium cyanide added to food during its preparation with intent to murder. However, the simple act of mixing cyanide into food is risky. If the food is acidic, such as foods containing vinegar (acetic acid) or fruits containing citric acid, the cyanide ions will react with the acid to release prussic acid vapour. The would-be poisoner might notice the innocent smell of almonds before he drops down dead. It is ironic that as the prussic acid is released from the food it depletes the amount of cyanide in the food to possibly a sub-lethal concentration. A case of the potential poisoner being killed by his own poison – rather like the terrorist who blows himself up in the process of assembling his bomb.

Prussic acid was planned as the murder weapon by Lizzie Borden to kill her father and stepmother in Fall River, Massachusetts in 1892. However, her efforts to purchase it were unsuccessful and she resorted to a brutal stabbing of the couple rather than the, more subtle, method of poisoning.

In Tokyo, Sadamichi Hirasawa entered the Teikoku Imperial Bank and pretended to be a doctor. He instructed the 14 bank employees to drink some of his medicine to avoid being infected by an outbreak of dysentery that was spreading throughout the local area. Each person was given a cupful of the medicine with the result that 13 of them died within minutes, allowing Hirasawa to help himself to the bank's takings. The surviving employee helped police in their efforts to catch the killer, who was then sentenced to life imprisonment. The 'medicine' that Hirasawa had prepared was a solution of potassium cyanide.

The story of Dr William Palmer's poisoning career includes prussic acid in addition to strychnine and antimony. Palmer qualified as a surgeon at St Bartholomew's Hospital in London in 1846. His poisoning activities had already begun while he was a medical student, when he added poison to his friends' drinks during drinking sessions, resulting in one of them dying. Palmer married a woman with a large inheritance and he took the decision to quit his medical career in favour of horse racing and gambling with his wife's money behind him.

He soon found himself financially stressed and when his mother-in-law fell ill, he decided to treat her in the expectation that if she died her money would go to his wife. It was not long before she met a sudden death, but Palmer was disappointed to find that no money went to his wife. His financial situation became worse and a succession of babies did not help matters.

It is said that Palmer poisoned the babies by dipping his finger in poison followed by sugar and then letting the babies suck his finger. Four out of his five children died. During their final minutes of life, they were said to have had convulsions. The financial problems worsened, and in an effort to raise some cash he took out insurance on the life of Anne his wife. She was soon dead and the insurance company paid up, but the amount Palmer received was insufficient to cover his debts. Next to be insured was his brother Walter.

Walter was a drunkard and was not expected to live much longer. However, Walter soldiered on, and Palmer became impatient and decided to give Walter a helping hand towards the grave. Palmer bought some prussic acid solution in Wolverhampton and within two days Walter was dead. However, the insurance company was suspicious and refused to pay up. Palmer then switched insurance companies and insured the life of a stable-hand that he was acquainted with. The second insurance company became suspicious and communicated with the previous insurer. Together they called in a detective, but while matters were being investigated, Palmer went on to kill a friend, John Parsons Cook, who died in agony in November 1855 with his back forced into the arched position.

A post mortem was performed on Cook to establish the cause of death, and Palmer was present during the cutting open of the body. At one stage, Palmer had access to the sample of stomach contents and somehow they got lost. Palmer was arrested for the murder of Cook.

The bodies of Anne and Walter were exhumed with the result that the analyst found antimony in Anne. At his trial, Palmer was convicted of the murder of Anne by means of antimony poisoning, Walter's murder by prussic acid poisoning, and Cook's murder by strychnine poisoning. Palmer was taken to

Stafford Jail where he was executed on 14th June 1856. He was the first person to be tried for murder caused by strychnine poisoning.

6.3 MAD MONK'S ORGY

This is another case of cyanide poisoning, but with some interesting aspects. The self-proclaimed Russian monk Grigori Rasputin (1869–1916), had gained influence with Tsar Nicholas II and his family in St Petersburg. Apparently Rasputin was allowed to treat the Tsarina's son who had haemophilia, a condition that causes massive bleeding even with the slightest injury, because of the blood's inability to clot. Rasputin gained even greater influence with the Tsarina, particularly when the Tsar was away during the World War I. This was seen by many as a threat to the power of the royal family and plans were made by people outside the royal family to murder Rasputin. The plot was to invite the monk to a party where he would be poisoned. The day of his poisoning was to be 16th December 1917.

An orgy with good wine and excellent food sounded tempting to Rasputin. The deciding factor was the chocolate cake on the menu, and he agreed to attend. Those conspiring to kill him laced the chocolate cake with potassium cyanide and slipped some into his wine just for good measure. Rasputin devoured the cake and quaffed the wine in large amounts, but the cyanide seemed to have no effect.

Desperate to see him dead, one of the conspirators shot him in the back but that also failed. He was then hit over the head and thrown into the river. When his body was found the authorities noted that, before he drowned, he had tried to get out of the water. Not only had he survived the cyanide, but also being shot and beaten. Why he was so resistant to cyanide is unknown, but it is claimed that he had a digestive condition that enabled cyanide to be broken down before it could do any harm.

6.4 AMERICAN GAS CHAMBER

In execution by means of cyanide we see something of a reversal from criminal poisonings to poisoning criminals. In 399 BC Socrates was executed by means of poison approved by the state, and in 20th Century America, execution was also by means of a state-approved poison. The Greek authorities sanctioned hemlock, whereas the American authorities opted for prussic acid.

America introduced the gas chamber as a means of capital punishment. Their first gas chamber execution took place in Nevada on 8th February 1924. The procedure was to firmly strap the condemned man into a chair beneath which was a compartment containing pellets of potassium cyanide. The executioner then left the prisoner and the airtight chamber was sealed. A tube led from the cyanide compartment to the outside of the gas chamber. Into this tube was poured sulphuric acid. As soon as the acid contacted the cyanide pellets, a fast chemical reaction took place and evolved a copious amount of prussic acid vapour. This diffused throughout the gas chamber to be inhaled by the convicted man.

The time taken for death varied from one execution to another, with one individual in agony for 11 minutes before death took him. During the execution, witnesses would watch the whole procedure through a window, so as to confirm that death had occurred. Following an execution, the prussic acid gas was neutralised by purging the chamber with ammonia gas.

There was much criticism of the gas chamber because of the intense suffering the condemned man was subjected to, and many states changed over to lethal injection for the death penalty. America abandoned the death penalty for a time but reintroduced it in 1976, with most states opting for lethal injection.

6.5 CYANIDE AND A SEDATIVE COCKTAIL

The biggest mass suicide occurred in Guyana during 1978 when 913 people were persuaded by Reverend James Jones to drink a cocktail of fruit juice, cyanide and sedatives. The people were members of a group known as the People's Temple and lived in a commune run by Jones in the appropriately named Jonestown.

Jones clearly had a powerful hold over his followers, but not all were sufficiently convinced as to commit themselves to taking the cocktail. This became evident when the scene was investigated and several bodies were found that had been killed by cyanide injection. Of all the deaths, many were suicides but those killed by injection were treated as murders.

6.6 CARBON MONOXIDE

This is a colourless, odourless poisonous gas that burns with a blue flame. Common sources are incomplete combustion from car exhausts and poorly maintained gas heaters. Carbon monoxide has no everyday use, but is consumed in large amounts in the chemical industry for synthesis.

When the concentration of carbon monoxide in the air is 0.01%, the symptoms are headache and nausea. At 10 times this percentage the symptoms become more serious: dizziness; confusion; convulsions; and finally respiratory arrest causes death. Delirium and strong visual experiences have been reported by those near to death by carbon monoxide poisoning. Some have reported seeing ghosts and believing they were in a haunted house.

In the past, carbon monoxide was a component of the coal gas that was used as the domestic gas supply. This explains why it was so frequently chosen as a means of suicide. The usual procedure was to place the head in an unlit gas oven. The American novelist, Sylvia Plath, who had been married to poet Ted Hughes, suffered with severe depression and ended it all by putting her head in the gas oven in 1963. Suicides by means of gassing using the public gas supply no longer occurs, as the gas is now methane (natural gas) which is not poisonous. Carbon monoxide suicides are often from inhalation of car exhaust gases and, increasingly, by lighting a portable BBQ in a confined space.

Carbon monoxide's poisonous action is rapid and comes from its ability to act as a haemoglobin blocker. Early symptoms of carbon monoxide poisoning are similar to being drunk, but blood tests readily distinguish between the carbon monoxide and alcohol, for in the blood carbon monoxide forms carboxyhaemoglobin. Evidence of carbon monoxide poisoning is the appearance of a cherry-pink colouration to the skin. We can compare this with the colour from cyanide poisoning, which is a deeper red.

6.7 TEN RILLINGTON PLACE

Carbon monoxide was the weapon of choice for John Reginald Christie who lured young women back to his house at 10

Rillington Place on the pretext of giving them medical treatment, or to relieve them of an unplanned pregnancy. He placed Friar's Balsam (a solution of plant extracts in alcohol) in a jam jar with a rubber tube connected to the mains gas supply, and closed by means of a spring clip. Another tube came from the jam jar and connected to a cardboard box.

The procedure involved the victim placing the cardboard box to her mouth while Christie released the spring clip allowing the gas to bubble through the balsam. Presumably the victim would not smell the gas because its odour would be masked by the strong smell of the balsam. The clinical smell of the balsam was no doubt chosen so as to fool the victim into thinking something medical was happening. Each inhalation exposed the victim to the balsam vapours laced with a hefty dose of poisonous carbon monoxide.

The box idea for delivering carbon monoxide with intent to kill was also tried by a Cranog Jones in 1991. He connected a square box to a flexible hose that passed through a hole in the house and into the garage. There the hose was connected the exhaust pipe of the Ford Fiesta. Before using it on his wife he tried it out on his daughter's pet cat and noted that it was dead within a few minutes. While his wife was sleeping he started the car engine, ran it on full choke and closed the garage doors.

The full choke would make the engine run fuel-rich and so ensure a high concentration of carbon monoxide in the exhaust gas. However, the closing of the garage doors would, in this instance, not increase the carbon monoxide level as the engine would not be aspirating the exhaust gases. Jones placed the box in his wife's breathing zone while she was asleep. He had overlooked the fact that his device acted like a sound box. Engine noise travelled along the hose and into the box. The noise woke Mrs Jones, who then called the police and ran off to hide behind an outside wall until they arrived. Jones was found guilty of attempted murder.

6.8 GREEN GAS CLOUD

Another poisonous gas is the green dense gas we know as 'chlorine'. It was discovered by Scheele in 1774 and shown by

Davy in 1810 to be an element. It is commercially important and manufactured on a large scale by electrolysis of molten sodium chloride. Chlorine has many uses, such as disinfecting water supplies, making the polymer PVC and producing chlorinated hydrocarbons that are important solvents.

The gas can be detected by its smell at as little as 0.2 ppm in the air. We are all familiar with its smell from municipal swimming baths, where it is added to kill harmful bacteria. At small concentrations of the gas in the air, it is irritating to the eyes and nose, but at higher concentrations it attacks the respiratory system. Many bleaches, such as *Domestos Original*, also use chlorine in the form of sodium hypochlorite. The disinfecting properties come from the powerful oxidising action of the chlorine, which kills bacteria by destroying their cell membranes. As a strong oxidising agent chlorine also breaks down many coloured organic molecules into simpler colourless molecules, hence its value as bleach.

There are many poisonous gases and vapours, and a proportion of them has been tried, since ancient times, as weapons against an enemy. The first large-scale use of a vapour was in World War I, when the Germans released chlorine gas, also known as 'bertholite', from thousands of cylinders along a four-mile front at Ypres on 22nd April 1915. The French soldiers on the Western Front were spellbound as a heavy green-coloured cloud rolled along the ground towards them. The gas poured into the trenches and filled them. It took only a few seconds after the initial exposure to the gas for the horrific effects of chlorine poisoning to be experienced. There was panic; the soldiers fought for air; there was frothing and gurgling in their throats; they became unconscious and choked to death.

Other chlorine attacks followed, but little military advantage was obtained because protective measures such as gas masks were soon introduced. Of course, wearing gas masks and cumbersome protective clothing seriously hampered a soldier's general mobility. In particular, the firing of a rifle was extremely difficult when a soldier was wearing a gas mask.

The poem, *Gas*, by war poet Wilfred Owen reveals the horror of a chlorine gas attack.

Dulce Et Decorum Est

Bent double, like old beggars under sacks,
Knock-kneed, coughing like hags, we cursed through sludge,
Till on the haunting flares we turned our backs,
And towards our distant rest began to trudge.
Men marched asleep. Many had lost their boots,
Drunk with fatigue; deaf even to the hoots
Of gas shells dropping softly behind.

Gas! Gas! Quick, boys! An ecstasy of fumbling
Fitting the clumsy helmets just in time,
But someone still was yelling out and stumbling
And flound'ring like a man in fire or lime –
Dim through the misty panes and thick green light,
As under a green sea, I saw him drowning.

In all my dreams before my helpless sight
He plunges at me, guttering, choking, drowning.

If in some smothering dreams, you too could pace
Behind the waggon that we flung him in,
And watch the white eyes writhing in his face,
His hanging face like a devil's sick of sin,
If you could hear, at every jolt, the blood
Come gargling from the froth-corrupted lungs
Obscene as cancer, bitter as the cud
Of vile, incurable sores on innocent tongues,
My friend, you would not tell with such high zest
To children ardent for some desperate glory,
The old lie: Dulce et decorum est
Pro patria mori

6.9 SARIN AND SOMAN

Despite attracting the condemnation of most of the World, the development of chemicals for mass poisoning went ahead with phosgene coming soon after chlorine. It is worth noting that the phosgene molecule is in fact a carbon monoxide molecule joined to a chlorine molecule. Phosgene is a colourless gas that is more poisonous than chlorine but, unlike chlorine, causes no irritation even when fatal doses are inhaled.

As the technology of mass poisoning by chemicals developed, three categories of chemicals were identified. 'Choking agents' included chlorine and phosgene. The latter was especially dangerous and accounted for the majority of deaths due to chemical attack. 'Nerve agents' included tabun, sarin, soman and VX. The first three are gases and are not persistent, so that when they come into contact with the victim some damage is caused, but soon they are carried off in the wind. VX, however, is a less volatile liquid, and once on the victim will persist and can cause death within 15 minutes. The nerve agents work by disrupting the chemistry at the nerve junctions with horrific consequences.

'Vesicants', or 'blister agents', were intended to incapacitate troops rather than kill them. The most feared vesicant was mustard gas (Yperite). In fact, it was a relatively involatile liquid, and it was this property that made it persistent. It was also chemically stable, so that it was not easily broken down and rendered harmless. Mustard gas had a delayed action of up to 12 hours after exposure before the symptoms of blindness, coughing, ulceration of the respiratory tract, and blistering of the skin appeared.

A big problem of using gases against the enemy is the wind. Having a weapon that relied upon something as unpredictable as the wind for its delivery, was bound to produce unexpected results. A change in wind direction shortly after release of the gas either dispersed it, without it ever touching the enemy or returned it to those who launched the attack. Despite the poor performance of warfare gases, the research continued. New chemicals were produced and improved ways of delivery such as placing the chemicals in shells, warheads and bombs, *etc.*, where they could be released by an explosive charge at the critical moment. A major development was the making of phosphorus-based organic compounds that behaved as nerve agents, and were effective at low concentrations. In fact, this was a spin-off from the German pesticide development programme.

Large stockpiles of chemical weapons were held by the Americans and Russians. Thankfully, the warfare gases were never used to a major extent. However, the nerve gases didn't go away altogether and make an occasional appearance in conflicts.

The riot control agent CS gas, often seen on news items being used by the police, is a chemical weapon, and can kill if the victim is in a confined space. The gas incapacitates by causing a burning sensation in the eyes, difficulty breathing, vomiting, and streaming from the eyes and nose.

Chemicals as weapons of mass destruction have an advantage in a warfare situation, in that they can incapacitate an enemy, but leave buildings and infrastructure intact.

6.10 BAD BUGS

In addition to chemical weapons that may be carried in the air, there are the biological weapons of germ warfare that also rely upon the air as a vehicle to take them to their target. Early attempts at using pathogenic bacteria as weapons were made in the Middle Ages. Diseased corpses were hurled into castles by means of catapults. We might wonder if those who placed the corpses upon the catapult prior to launching, also suffered from the bugs. Perhaps there were sacrificial people employed for the task. As with the gases in chemical warfare, it is always likely that the wind will change direction, giving the aggressor a taste of his own medicine. Whenever there are interests in weapons and warfare there is secrecy, suspicion and self-interest.

It is thought that some laboratories still carry out research aimed at the developing of biological weapons. With advances in genetic engineering it is almost certain that some countries are active in this type of research. What goes on under the guise of 'research for defensive tools' is adaptable to development of offensive weapons.

Some areas in which it is likely that research continues, are in aerosol droplets to deliver nasties such as smallpox, pneumonic plague, anthrax and glanders. Biological weapons can persist in the environment for a long time. For instance, in World War II, trials with anthrax bacteria were performed on Gruinard Island, which remained an anthrax risk for the next 56 years.

Germ warfare relies upon biological toxins, which are carried to the enemy by means of bacteria, viruses or fungi. These microorganisms can be used to attack an enemy through its food, its water supply, or in the air it breathes. Once they have reached their target, the biological weapons can both infect and reproduce to spread disease even further. These weapons can be developed to kill crops, people or animals. As with chemical weapons, the buildings and infrastructure are left undamaged.

6.11 WOOLSORTER'S DISEASE

This is an acute disease caused by the anthrax bacterium. Normally confined to farm animals, the bacterium can be carried to humans by contact with hair, hides or excrement. Woolsorter's disease is due to the bacteria being transmitted during the handling of wool and pelts. The disease can attack the lungs causing pneumonia, or the skin producing malignant pustules. In the past, anthrax killed on a large scale until Louis Pasteur developed the first vaccine against the disease in 1881. Nowadays, anthrax can be treated by giving the patient large doses of penicillin or tetracycline. With an understanding of the problem, improvements in public health and better practices, the problem is no longer a serious threat. The main risk of infection is when animal materials are being processed to make saleable products.

Anthrax is a soil-borne spore that can survive harsh conditions for many decades. When dust from the soil is inhaled, the spores are absorbed into the lungs where they become activated and begin to multiply as bacteria, causing pneumonia. If the spores are in contact with broken skin, ulcers are produced.

There are many strains of anthrax, with the most powerful being tested on Gruinard Island in Scotland by the British military scientists from Porton Down in 1942. At that time, there was a lot of interest in developing biological weapons for use in the war. The tests on Gruinard involved a bomb containing anthrax spores being exploded in a field of sheep. After only three days, the sheep began to die. The island was so badly contaminated with anthrax that it was declared out of bounds for over 50 years.

In 2001, anthrax spores were used in a bio-terrorism attack in American. Letters addressed to Democratic Senators were posted. These contained anthrax spores in a form that would easily form a mist when the letters were opened, thereby allowing the spores to spread through the air for people to unknowingly inhale. Five people died in the attack and several were made seriously ill. Inhaling anthrax spores starts off with mild symptoms such as fever, fatigue, malaise, coughing and a feeling of pressure on the chest. Then a second set of symptoms starts and includes internal bleeding, blood poisoning and meningitis. Within about seven days the patient is dead.

6.12 PHOSPHORUS

The element phosphorus occurs in a few different forms but it is the one known as 'white phosphorus' (sometimes called 'yellow phosphorus') that is treacherous and the type used to make some terrifying weapons such as incendiary bombs, shells, rockets, grenades, and in making smoke screens. One form of incendiary device was known as Fernian Fire. It comprised a glass bottle that was filled with phosphorus dissolved in carbon disulphide and used like a Molotov cocktail. The incendiary weapons caused some horrific burns even when only small amounts of burning phosphorus came into contact with skin.

White phosphorus is a waxy solid that is stored in water. Once out of the water it reacts with oxygen in the air, evolving a dense white smoke of phosphorus pentoxide. Phosphorus may be seen in the dark as the 'glow of phosphorus' due to this reaction. The element does not occur in nature because of it being highly reactive, and must therefore be obtained from those phosphorus compounds that occur in nature such as phosphate rock.

The first preparation of white phosphorus was by Hennig Brand, in 1669. He discovered phosphorus during an alchemical experiment, in which he was trying to make gold by distilling fermented urine. In modern times, phosphorus is manufactured by chemical reduction of phosphate rock using coke. As a rodenticide white phosphorus played an important role, but has now given way to modern poisons. White phosphorus was mixed with moist flour to form a paste of about 3% phosphorus.

Phosphorus poisoning occurs when someone swallows phosphorus, often as rat poison, in attempting suicide. Even small amounts of ingested white phosphorus are serious, as severe irritation of the gastro-intestinal tract occurs followed by bloody diarrhoea, liver damage, skin eruptions, circulatory collapse, coma, convulsions and death. In the early stages there is a smell of garlic on the breath.

6.13 ACIDIC ATTACK

Unlike most poisons that are carried around the body to a target organ, the corrosive poisons are indiscriminate. They work by severely damaging whatever living tissue they come into contact with. We can think of them as being brute-force poisons. Common examples are the liquids: bleach; caustic soda solution; and sulphuric acid. In the early 20th Century these were used for poisoning, because there were restrictions upon traditional poisons and, unlike modern times, there were few drugs available for poisoning by overdose.

Most poisonings by corrosive chemicals are, these days, accidents. They often involve children drinking household chemicals. It is hard to imagine how any of them could be administered for poisoning. Symptoms of swallowing corrosive poison are severe damage to soft tissue in the mouth, oesophagus and stomach. There is also difficulty in breathing, speaking and swallowing.

In the late 19th Century we find examples of acids being used for poisoning. The first involved Elizabeth Berry. Her husband, mother and son died over a period of years and Berry collected the insurance. There were no suspicions until the day when Berry decided to poison her 11-year-old daughter, Edith Annie. The poison chosen was a highly corrosive concoction of creosote and sulphuric acid, which the child was forced to swallow. The attending doctor noted burns and stains around the dead girl's mouth and a death certificate was refused. Elizabeth Berry was hanged at Liverpool in 1887 for the murder of her daughter.

Another poisoning involving the death of an 11-year-old daughter and a corrosive chemical, involved William Dawson Holgate, a coal merchant in Bradford. His daughter, Lily, was insured for £20. To the dismay of many, Holgate was acquitted when it was reported that Lily could have taken the poison by mistake. However, as the poison was phenol, a chemical similar to the creosote used in the Elizabeth Berry case, accidental poisoning seems most unlikely. Even the tiniest amount in the mouth would be instantly spat out before anything near a lethal dose could be swallowed.

Another well-known case was that of Martin Slack who gave his baby daughter a drink of nitric acid. The baby's mother, who was in another room, heard the baby scream and dashed in to find yellow liquid oozing from her mouth and which burnt the mother's fingers. The baby was dead within the hour. Slack's motive was to avoid paying the weekly five-shilling maintenance to the baby's mother.

Today, hospital emergency departments frequently encounter poisoning by accidental ingestion of corrosive chemicals. The general procedure, for strong acids, is to give the patient large quantities of milk or egg whites and/or use gastric lavage. If the acid can be diluted, then its corrosive power is largely destroyed. To appreciate how destructive corrosive chemicals can be to human flesh we should examine the case of the acid bath murders carried out by John George Haigh in the 1940s. His victims were killed by conventional means, but the disposal of their bodies was far from conventional. Haigh immersed his victims in sulphuric acid, which caused almost total disintegration within a matter of days.

Sulphuric acid in contact with flesh performs a variety of chemical reactions that destroy the flesh. First there is physical dehydration of tissue due to sulphuric acid's power to absorb water. The reaction produces heat, which helps speed up other destructive reactions occurring simultaneously such as the breakdown of fats and proteins. The fats are converted into fatty acids and glycerol, whereas the proteins form amino acids. These three products are water-soluble and dissolve in the acid that has become diluted due to it having absorbed water from the body. Other chemicals undergo reactions that produce a black residue of carbon.

Bones are made up of calcium phosphate and the protein collagen. Reaction with sulphuric acid converts the calcium into calcium sulphate and the phosphate is changed into phosphoric acid. This is a slow process because the calcium sulphate formed is insoluble and only slowly disperses. The collagen breaks down and adds to the other carbon residues. Other reactions run in parallel and in series with these main reactions. By such chemistry, sulphuric acid converts the human body into a thick evil-smelling black sludge. It was this sludge that the police had the pleasure of discovering when they investigated Haigh's premises.

6.14 PIMLICO POISONING

Chlorinated hydrocarbons are very good solvents for grease, and most of them are non-flammable. These properties make them highly desirable for industrial cleaning purposes, especially degreasing of electronic components and dry cleaning of clothes. These solvents do not occur in nature and must be synthesised by reacting chlorine with hydrocarbons from crude oil. Chlorinated hydrocarbons also have anaesthetic properties when their vapours are inhaled. Chloroform is the best known example, being introduced into hospitals in the 1840s.

Chloroform featured in the Pimlico mystery of 1886 when Thomas Edwin Bartlett was found dead. A post mortem revealed that he had chloroform in his stomach, but there were no signs of damage to the throat and windpipe. This baffled those involved in the post mortem, as the swallowing of chloroform always produced throat burns. However, Thomas' wife Adelaide was arrested under suspicion of poisoning him. There were many who believed Thomas had committed suicide because of the suffering in his life. He had tapeworms, rotten teeth and his wife was over-friendly with another man. Whether suicide or murder, the mystery of how the chloroform got into his stomach remained. Adelaide was acquitted. The chloroform had been obtained on a prescription issued by Dr Alfred Leach but it is curious that the four small bottles prescribed were collected from different chemists.

Another one of this group of chemicals to find medical use was trichloroethylene, which made a difficult childbirth easier. Unfortunately, the dosage for inhalation of the vapours from these chemicals was dangerously near to the fatal dosage. However, the discovery of chloroform's anaesthetic properties led the way to safer compounds, such as the halothanes we now have as modern anaesthetics.

Another chlorinated hydrocarbon is carbon tetrachloride. This was in common use for cleaning electrical contacts, as a degreasing agent, and for removal of stains from textiles. The father of comedy actor Kenneth Williams died after accidentally drinking the solvent, which was in a cough medicine bottle. The mistake is understandable as some cough medicines of the time contained chloroform, which has a similar odour and causes a burning sensation in the mouth.

6.15 GLUE SNIFFERS

Chlorinated hydrocarbons were available in household products such as adhesive for resoling shoes, in correction fluid such as the original Tippex, and in paint stripper. Domestic use of chlorinated hydrocarbons was phased out to reduce the atmospheric pollution that damages the ozone layer, and because children were inhaling the narcotic vapours with fatal results.

Chronic poisoning occurs where there has been long-term exposure to chlorinated hydrocarbons, because they easily dissolve in the body fats. Solvent sniffing activities have now switched from chlorinated hydrocarbons to aerosol propellants, and other volatile compounds capable of a psychoactive effect. But these are still dangerous and tragically youngsters continue to accidentally poison themselves. Poisoning of the atmosphere by chlorinated hydrocarbons is now less than it was a few years ago, but we are still left with a significant hole in the ozone layer, which is implicated in the rise in skin cancer.

6.16 STALIN'S WARFARIN

Humankind has always been faced with the problem of preventing insects and rodents from consuming or otherwise damaging food supplies. The main solution to the problem has been to use different chemicals to poison the offending creatures. Nature has its own range of pesticides, such as the insecticide pyrethrum that the chrysanthemum produces for protection against insect attack. Pyrethrum is used on crops, but the amounts available from nature are limited. Obtaining pyrethrum was achieved by grinding and solvent extraction of the chrysanthemum flowers.

Nowadays man-made pesticides are manufactured and they have almost completely replaced natural pesticides. Production is on a massive scale with two million tonnes a year being made to satisfy the needs of modern intensive agriculture. It wasn't until the development of the organic chemical industry that synthetic chemicals became available – the start of man-made pesticides. It began in 1939 when the insecticidal properties of DDT were discovered. Other synthetics were soon invented, and the man-made pesticides industry snowballed, and with it came a mountain of environmental problems.

For instance, DDT was effective in killing crop-destroying insects, but it caused havoc to many species of birds that used those insects as their main food supply. The reporting of this, along with the publication of Rachel Carson's *Silent Spring* made us sit up and think about the consequences of poisoning living systems on a large scale. Now we have an understanding of the food chain and the destiny of poisons that enter it. The arrival of highly sensitive analytical instruments to perform pesticide residue analysis was in the nick of time.

For killing weeds, paraquat is now widely used. This contact herbicide, if swallowed, causes severe lung damage. It destroys the tissue within a few days and is nearly always fatal. Overall, the symptoms of paraquat poisoning are ulceration of the digestive tract, diarrhoea, vomiting, kidney damage, jaundice, haemorrhage and fibrosis of lung tissue, leading to death from lack of oxygen.

Warfarin is a relatively recent poison, being introduced as a rodenticide in the 1940s. It is synthesised from the coumarin that is present in woodruff, a plant that was found to be responsible for haemorrhaging in cattle after they had been de-horned. An odourless and tasteless compound, warfarin interferes with the coagulation mechanism that thickens the blood, and prevents continued bleeding from a wound. Due to this anticoagulant effect, warfarin is prescribed to those who have had a heart attack in order to help prevent a recurrence. Stalin (1953) bled to death and it is believed that he was poisoned with warfarin. Large doses of Vitamin K may have saved him if administered in time, as this is the blood-clotting vitamin and is used as the antidote to warfarin.

6.17 DETERGENTS, HORMONES AND PLASTIC

Traditional soap was made by reacting sodium hydroxide solution (caustic lye) with animal fats or vegetable oils. When used for washing and the waste water discharged, the soap underwent bio-degradation because the main part of it, the fatty acid, was a natural substance. It was soon broken down and rendered harmless.

With the arrival of synthetic detergents for domestic use in the 1950s, there began some environmental problems. Many rivers began to produce huge volumes of foam because the detergents in the sewage effluents did not bio-degrade, and so persisted in the environment. Later ethoxylate detergents were introduced and these also brought the problem of persisting in the rivers and oceans. Some of the ethoxylates mimicked human hormones, and concerns grew about men developing breasts if they ate fish from waters contaminated with ethoxylates, the 'gender bender' chemicals. This environmental poisoning due to man-made chemicals was a driving force in the study of environment and ecosystem changes.

Plastic problems are becoming worse due to material physically breaking down into small particles as they are churned around in the oceans. We use many plastics that do not biodegrade, but most of these will break down due to the effect of ultraviolet light from the sun in combination with an oxidising effect from the air. But plastics submerged in the oceans are away from these influences and are stable, for how long, no-one knows. This is relatively new chemistry. One thing is sure and that is the problem is getting worse. As the particle size is reduced, the plastics are finding their way into living systems.

At a particle size of a few microns, they can move around within marine life. A worry is that the reduction of particle size may take the plastics down to almost nanoparticles. To aggravate the problem, these synthetic polymers are lipophilic, that is, they are attracted to chemical pollutants that are not dissolved in the sea water but are simply dispersed. The tiny plastic particles mop up these chemicals, concentrate them and then convey them into living systems.

Chemicals that would otherwise be unable to become part of the living system are then able to piggy-back their way into the food chain. For example, organic mercury compounds can be formed when mercury-polluted discharges enter the sea, as occurred at Minamata Bay. These organic mercury compounds are readily absorbed by the plastic particles.

6.18 CHINESE MELAMINE

Melamine is manufactured on a large scale for manufacturing a plastic known as 'melamine formaldehyde'. The plastic is widely used for coating chipboard to produce the melamine laminate for the flat-pack wardrobes and bookshelves that are popular at the budget end of the furniture market.

In the 1990s, Chinese production of melamine grew dramatically and a surplus was generated. With insufficient capacity to convert this to the plastic, some enterprising individuals turned their minds to thinking up new applications. Melamine is a relatively tasteless white powder that contains a high percentage of nitrogen. This nitrogen content was the key to melamine's new application, in which it was to be added to powdered baby milk.

The quality of milk is rated on the basis of how much protein it has. This involves analysing the milk powder to measure its total nitrogen content. All protein has a certain level of nitrogen, and once the nitrogen content of the sample has been determined, a simple calculation then gives the amount of protein. Adding melamine, with its high nitrogen content, to milk powder increases the figure for the analysed nitrogen and the subsequent calculation shows a high, but false, protein level. Thus, poor-quality milk powder that was low in protein was being adulterated with cheap melamine to give the impression of the milk powder being of high protein and therefore of high quality. This made it suitable for the baby-milk market.

The perpetrators of this deception had taken no account of the poisonous nature of melamine, or perhaps they had simply ignored it, as the prospect of easy money was an opportunity not to be missed. In the body, melamine reacts with cyanuric acid in the bloodstream and forms yellow insoluble crystals that block the kidneys and stop them working. The result was that 53 000 children were made ill in what became known as China's Milk Scandal. In January 2009, two people were found guilty of the offence and sentenced to death.

6.19 SCREENWASH AND ANTIFREEZE

Present in car screenwash and methylated spirit, methanol has caused many accidental poisonings. For example, methanol poisonings occurred during America's prohibition of alcohol due to the drinking of spirits that were produced by improper distillation practices. In Bombay in 1992, illegal spirit killed many people during celebrations.

The antifreeze added to the cooling water in a car radiator is usually ethylene glycol, with a bit of blue dye. The pleasant, sweet taste and colourless appearance of pure ethylene glycol explained why it was chosen to doctor white wine, a practice that had disastrous consequences. It is not the ethylene glycol that is poisonous, but the oxalic acid that is formed when ethylene glycol is oxidised in the liver. When the oxalic acid arrives in the kidneys, it forms clusters of calcium oxalate crystals that block the filtering action of the organ. The result is kidney failure.

Lynn Turner used antifreeze to poison her husband Glen Turner, and then her lover Randy Thompson. The post mortem of Thompson's body revealed calcium oxalate crystals in his kidneys. In May 2004 Lynn Turner was convicted of the murder of Randy Thompson and given a life sentence.

Jacqueline Patrick was given a 15-year sentence for attempted murder. She spiked her husband's drink of Lambrini with antifreeze whilst they were having a drink at their London home in 2013. Jacqueline's daughter, who had encouraged the poisoning, also received a three-year sentence.

6.20 PHOSSY JAW

The capacity of white phosphorus to burn when in contact with air was the basis of the first strike-anywhere matches. Workers in the match industry between the mid 19th and early 20th Centuries, particularly the girls who dipped the wooden sticks into the phosphorus paste, were exposed to phosphorus fumes. Over a number of years of exposure, the employees developed a condition known as 'phossy jaw'.

This was due to phosphorus attacking the lower jawbone, and resulted in an abscess in the skin around the jaw from which evil smelling puss oozed. Prior to the development of phossy jaw the match workers also suffered brain damage, painful toothache and swelling gums. It is said that bones in which phosphorus had accumulated glow with a greenish white light. Concern over the appalling damage to health provoked anger among the employees, that led to the London match-girls' strike of 1888.

Oral ingestion of white phosphorus results in damage to the liver, heart and kidneys. It also produces 'smoking stool syndrome' where the faeces produce white smoke – and possibly glow in the dark. Phosphorus was until recently popular as a rat poison, and could be purchased as such from the local chemist.

This was how Mary Ansell, a domestic servant, obtained the poison. She baked a cake with phosphorus in it and sent some to her sister, Caroline who was in Watford Mental Asylum. Caroline shared the cake among fellow patients but ate most of it herself. Within hours they became seriously ill and Caroline died four days later. Mary's motive was to collect the insurance of £22 on her sister's death, but suspicions were aroused, a post mortem was held and Mary was arrested. After being found guilty, she was hanged on 19th July 1899 by James Billington outside St Albans Jail and buried within the grounds of the jail.

6.21 NITRO NASTIES

The nitro chemicals are important as explosives, and therefore find use as military weapons. Thus, we have nitrocellulose as guncotton, trinitrotoluene, picric acid, TNT and nitroglycerine. During the war years, the people working on the production of these nitrochemicals noticed that they lost weight without changing their diet or increasing exercise.

The reason is that many of these nitro chemicals when absorbed by the human body increase its metabolic rate, which causes calories to be burnt off. This may good news for those with a weight problem, but these chemicals are lethal. The idea was picked up on in the 1930s when diet pills containing nitro chemicals as the active ingredient were marketed. In 1934 a young dancer intent on losing weight started on a course of dinitrocresol, which involved taking one capsule daily. But in her eagerness to lose weight she had exceeded to recommended dosage, which proved fatal. A compound similar to dinitrocresol is dinitrophenol (DNP). The latter increased in popularity, but the pills were soon banned when the dangers were recognised.

In 2013 in Britain, Chris Mapletoft, a gifted young rugby player, died after buying DNP pills online. Also in 2013, Sarah Houston, a Leeds University medical student died from DNP poisoning after taking pills bought on the internet. A few years before, 26-year-old Selena Walrond had died after taking a large dose of DNP in a desperate effort to lose weight. Two years later, Eloise Aimee Parry, a 21-year-old British student, died within a few hours of taking diet pills bought over the internet. The pills contained DNP. The Coroner reporting on the case warned the public against purchasing medicines or supplements from the internet.

The symptoms of DNP poisoning include an increase in temperature, profuse sweating, nausea, vomiting, collapse and death. DNP in its pure form can also cause poisoning by absorption through the skin and inhalation of its vapour.

6.22 AGENT ORANGE

There are several chemically related dioxins, but the one referred to simply as 'dioxin is in fact TCDD'. We must be careful not to confuse this with the chemical called 'dioxan' that is used as a solvent. TCDD is formed as an impurity in the production of the herbicide 2,4,5-T, especially if the conditions are not tightly controlled. This herbicide, known as Agent Orange, was used on a large scale by American forces in Vietnam to remove forest cover and the enemy's crops, between the years 1962 to 1971. Unfortunately, the Agent Orange used had higher than normal levels of the impurity dioxin.

It is claimed by some war veterans that they suffered ill health as a result of their exposure to dioxin when Agent Orange was being handled and sprayed. Later, it was found that about 40 000 000 L of the contaminated product was sprayed onto 12% of the area of South Vietnam. TCDD is a stable compound, and this enables it to persist in the soil for decades in the areas where the Agent Orange was sprayed. After the war, many areas of contaminated ground were studied, and blood samples from the people who lived in the affected areas were analysed. The result was that the blood TCDD levels were much higher than average, because these people were growing their food crops in the poisoned soil. They were also taking their drinking water from contaminated sources. The authorities in Vietnam blame TCDD poisoning for the higher than normal levels of birth defects, and mental and physical disabilities.

There are many dioxins and most are toxic, with TCDD, the most poisonous, being the focus of much toxicity research. Other cases of accidental exposure to dioxin are recorded, with the worst being the accidental release of chemicals at a works in

Seveso, Italy in 1976. A reaction vessel in the chemical works over-heated, built up excess pressure and released a cloud of gas and dust which settled on the Seveso area with a population of 17 000.

Dioxin can also be formed when waste PVC is incinerated and then discharged into the air. One symptom of dioxin poisoning is chloracne, a very severe form of acne. Dioxin was thought to have been used in an attempt in 2004 to poison the Ukrainian opposition candidate Viktor Yushchenko. He developed chloracne and dioxin was detected in his blood.

6.23 MAD HATTERS

Many other cases of chronic poisoning from industrial processes are documented. Before the days of occupational safety, dental technicians were prone to mercury poisoning when they inhaled mercury vapour while preparing the amalgam from liquid mercury.

Some of the worst mercury poisoning cases came from the hat industry in Britain during the 19th Century, when mercuric nitrate was used as part of the process of making felt. This involved soaking the skin of small animals in the mercury solution to separate the fur, which then would be further processed to form the felt. The drying process was especially risky as mercury would have been released as vapour into the air and subsequently inhaled. And the handling of the dried fur would likely have released dust containing mercury.

Inhalation of the toxic vapour and dust led to chronic mercury poisoning. The symptoms of this were due to the mercury attacking the central nervous system causing tremors, irritability and mental instability. This gave rise to the term, 'mad hatter's disease' and a recognition of the occupational poisoning from mercury. The mad hatters are referred to in Lewis Carroll's book *Alice in Wonderland*, which was published in the 1860s.

6.24 CHIMNEYS AND CHILD CANCER

As the industrial revolution developed throughout Britain, the number of domestic coal fires increased. During use the insides of the chimneys became coated with soot. This was simply unburnt carbon from the coal, but also contained creosote, which was a mixture of phenols that distilled out of the coal. The combination of soot and creosote formed an adherent layer inside the chimney, and this had to be periodically swept out. Failure to remove the sooty deposit could lead to chimney fires and even house fires due to the flammability of the soot and creosote mixture.

Domestic chimneys were too small for an adult to climb up with a brush, and so children were employed to climb up inside the flue and brush away the soot. In Britain, children as young as four were apprenticed to the job. They were in effect slaves and were legally the property of the master sweeper with whom they lived. The domestic arrangements were appalling, and children had to sleep under coal sacks as blankets and with no proper access to washing facilities. Most were boys, but a few girls were also employed. Daylong contact with the soot led to health problems.

Many boys in their teens and early twenties who had worked as climbing boys when they were small, developed cancer of the scrotum. When the facts were studied it was concluded that the contact with the soot was the cause. This was the first recorded cancer related to coal burning. A modern chemical analysis of the sooty deposit in the flue shows that in addition to creosote and soot, there is also fine ash lifted from the hearth by the up-draught. This ash contains a large number of poisonous metals as their oxides. For example, a typical coal ash analysis shows the presence of arsenic, beryllium, cadmium, chromium, lead, thallium and vanadium.

It is worth noting that much of the fine ash, known as 'fly ash', went into the atmosphere to join the ash from the many mill chimneys. The problem of soot and ash was therefore not confined to the profession of chimney sweeping, but also affected the population as a whole.

6.25 CANARIES, CANDLES AND COAL

Coalmining has been responsible for many poisonings. A big problem in mines is due to the fact that they are enclosed. There is little, if any, natural ventilation and so gases can build up. This is less of a problem in modern mines as they use forced ventilation and are supplied with fresh air. We might think that coal is an inert mineral that has been there for millions of years. However, in its rock-like structure there are gases. Some are from ancient times and some are being produced. Cutting into coal seams often releases these gases or allows water to access reactive minerals to produce additional gases.

To understand this, we might consider the chemistry of coal formation. Coal measures formed when trees and plants decayed in ancient swamps. The decomposing remains formed sediments that eventually became covered with other sediments. Thus, the decayed plant material, which started as carbohydrate, became compressed and heated as it was forced deeper into the Earth's surface. During the millions of years, the decayed material released gases as it became carbonised into coal. Of course most gas was lost as it seeped out through surrounding rocks, but some was retained. The main gas is methane, but there may also be carbon dioxide, carbon monoxide and hydrogen sulphide. The latter is formed when sulphides react with water that gets in when a coal seam is split.

Methane and carbon monoxide form an explosive mixture in air. Carbon monoxide and hydrogen sulphide are deadly poisonous. When a coal is mined, gas is released. Coalminers are very much aware of the dangers of working there, and there is a tragic history with many deaths from explosion, fire and poisoning.

The table summarises these and lists the terms used by the miners. The word 'damp' comes from the German word Damf meaning vapour (Table 6.1).

Table 6.1 Gases found in coal mines.

Mining name	Gases present	Notes
Blackdamp	Carbon dioxide and nitrogen	Oxygen from air has been removed by corrosion with iron parts in mine. Risk of suffocation. Candle flame extinguished. Canary dead.
Afterdamp	Nitrogen, carbon dioxide and carbon monoxide	Formed after a mine fire or explosion. Poisonous and suffocating.
Firedamp	Methane	Explosion risk.
Stinkdamp	Hydrogen sulphide	Smell of rotten eggs. Deadly poisonous and risk of explosion.
Whitedamp	Carbon monoxide	Odourless. Deadly poisonous gas and risk of explosion.

In addition to the risk of fire, explosion and poisoning from gases, the coal dust was also a problem. When it became extremely fine and dispersed in the air it was explosive. Furthermore, when inhaled by the miner it caused a disease known as 'black lung'. Another lung condition caused by inhalation of dust was silicosis caused by fine particles of silica becoming embedded in the lining of the lung. Asbestos workers also suffered in a similar way, due to inhalation of tiny fibres of asbestos, which is also a form of silica. Although not strictly speaking poisons, these dusts were just as bad in the amount of disease and premature deaths that they caused.

As coalmining developed, the need for air monitoring was accepted. At first a candle flame was used. By noting how it burnt, the miner could assess the quality of the air. Of course the presence of a naked flame was itself hazardous. Canaries were also used to check the quality of the air. They have a very high aerobic rate and soon respond to poisonous gas or low oxygen levels. A major improvement was when Humphrey Davy invented the miners' safety lamp. In modern times there is constant monitoring of the air quality in the coalmine using infrared or electrochemical detectors to measure the level of poisonous gases, and to check the oxygen content of the air.

We should spare a thought or two for the early 19th Century coalminers. As the industrial revolution became more productive it demanded raw materials at an increasing rate. Water wheels were being replaced by steam engines, increasing the demand for coal. There were coal seams near to the surface and some were exposed at outcrops. The coal layers of interest were around 14–24 inches thick. Being too small for adult miners, children were sent to burrow below. There were no safety standards. Many children in Britain's industrial revolution died by gas poisoning. Others were caught in explosions, trapped, crushed to death and sometimes left there as dead.

6.26 RADIOACTIVE GIRLS

Luminous dials for watches and aircraft instruments had the hands and numbers painted with radium paint. This was usually done by young women who would obtain a fine point of their paintbrushes by licking the bristles of the camel hair brushes.

Some of the girls went one step further and not only painted the watch hands, but also the fingernails of their own hands. Presumably they wanted to see what their hands were doing in the dark.

The exposure from licking the brushes caused a serious disease of the jaw bone known as 'radium jaw', which was similar to the phossy jaw that the women in the match industry suffered from. In 1925 a group of these women, known as the Radium Girls, sued their employer, US Radium Corporation, New Jersey, for causing occupational disease due to radiation poisoning. This case was a milestone in America in moving towards legislation relating to occupational diseases.

In the court case it emerged that the girls had been led to believe that the paint was harmless. The company's managers and chemists made all efforts to avoid direct contact with it, by handling it with tongs while wearing masks. It was also stored behind lead screen so as to minimise the amount of radiation that the chemists were exposed to. Furthermore, the company had produced literature on the dangers to health when using the radium paint. It seems there was a cover-up by US Radium and other companies involved in the radium paint business, and evidence relating to the dangers was suppressed.

The use of radium paint for watch and instrument dials ceased in the 1960s. However, there are still many such dials in use. They may no longer be luminous but will still be radioactive, as the half-life for the radium isotope is 1600 years. The radium in the luminous paint comes from the radioactive decay of uranium found in minerals such as pitchblende. Loss of luminosity occurs because of the breakdown of the zinc sulphide, which is the chemical that converts the radiation into visible light.

Radium Girls **by Eleanor Swanson**

We sat at long tables side by side in a big
dusty room where we laughed and carried
on until they told us to pipe down and paint.
The running joke was how we glowed,
the handkerchiefs we sneezed into lighting
up our purses when we opened them at night,

our lips and nails, painted for our boyfriends
as a lark, simmering white as ash in a dark room.
"Would you die for science?" the reporter asked us,
Edna and me, the main ones in the papers.
Science? We mixed up glue, water and radium
powder into a glowing greenish white paint
and painted watch dials with a little
brush, one number after another, taking
one dial after another, all day long,
from the racks sitting next to our chairs.
After a few strokes, the brush lost its shape,
and our bosses told us to point it with
our lips. Was that science?
I quit the watch factory to work in a bank
and thought I'd gotten class, more money,
a better life, until I lost a tooth in back
And two in front and my jaw filled up with sores.
We sued: Edna, Katherine, Quinta, Larice and me,
but when we got to court, not one of us
could raise our arms to take the oath.
My teeth were gone by then. "Pretty Grace
Fryer," they called me in the papers.
All of us were dying.
We heard the scientist in France, Marie
Curie, could not believe "the manner
in which we worked" and how we tasted
that pretty paint a hundred times a day.
Now, even our crumbling bones
will glow forever in the black earth.

(With permission from Eleanor Swanson)

6.27 RADIATION POISONING

Marie Curie suffered from radiation poisoning resulting from her work on radium, and possibly earlier exposure to X-rays when she worked on radiography units during World War I. In her radium research she extracted the element from pitchblende, a rock containing uranium. The radiation to which she was exposed induced a disease in her bone marrow that eventually killed her.

Radiation sickness comes from exposure to ionising radiation, and may be either acute or chronic just like many other types of poisoning. Radiation is all around us and is mostly from natural sources, although there is a growing level from man-made sources. The natural background radiation comes from the rocks in the Earth and from outer space. Industrial and medical sources of radiation are from radioactive isotopes that are increasingly being used. And radioactive isotopes enter the environment from accidents and leakages from nuclear power facilities and nuclear weapons. A growing source of radiation is from orphan materials and hardware from redundant processes. Occasionally some of these go missing, which is a growing concern. Whose hands have they fallen into and for what purpose?

The main danger with radioactive materials is the radiation they release. For example, the thyroid gland absorbs iodine to produce thyroxine. The iodine sources found naturally have negligible radioactivity. But the iodine emitted during testing of nuclear weapons or from accidents with nuclear reactors is radioactive iodine, which the thyroid converts to thyroxine that releases radiation, which damages the tissue.

So long as we keep a safe distance from radioactive sources, or keep them stored in special containers there is no problem. But if we are exposed to them and absorb them, this can be a serious risk to health. Absorption may be either the isotope itself or a material that it is a part of. When this occurs, the substance may remain in the body, perhaps fixed into a target organ, and continue to release damaging radiation.

Among the rocky material, and not far beneath us, uranium is undergoing radioactive decay and releasing decay products. The speed at which uranium decays is measured by its half-life which, for most of the uranium on Earth is five billion years. This means that it takes five billion years for half of the uranium to decay. As such, the uranium in the ground is decaying and, in that process, releasing radon gas, which itself is radioactive and emits alpha radiation.

The radon eventually reaches the surface and disappears into the atmosphere causing no harm to anyone. However, buildings prevent the gas escaping and allow it to accumulate. The gas is dense, colourless and odourless. It builds up in basements and we have no way of knowing it is there, unless a radon monitoring device is installed. For most buildings and occupants, the amount of radiation is so small as not to pose a risk to health. However, in some areas the radon can reach concerning levels.

Radon risk areas show a random distribution. Locations with large amounts of granite in the ground are more at risk than most areas. It is interesting that the concentration of radon in houses shows seasonal fluctuations, with a minimum in summer and maximum in winter. This is nothing to do with the rate of radioactive decay, but is due to people opening the windows in summer and closing them for winter.

The greatest risk of radon exposure is in mines, especially if they have poor ventilation. Even in the 16th Century it was reported that miners suffered a wasting disease of the lung. This, we now know to be lung cancer. As we learn more about radiation exposure, we know that it is alpha radiation that causes the problem when the radon gas builds up. When the gas is inhaled, the radiation emitted is absorbed by the surrounding tissue where it damages the DNA molecule. This results in the cell mutation that leads to a tumour.

Radon represents our greatest day-to-day exposure to radiation, but it is not the only source. We are exposed to cosmic rays that have been travelling through the cosmos for billions of years and gamma rays are released from the ground and buildings (Table 6.2).

Table 6.2 Examples of radiation sources and uses.

Americium-241: alpha emitter, domestic smoke detectors
Radium-226: alpha and gamma emitter, luminous paint on watch dials and aircraft instrument dials
Radium-223: alpha emitter, injection to kill cancer cells
Cobalt-60: gamma emitter, sterilising food, destroying tumours
X-Rays: produced in a discharge tube for radiography
Iodine-131: beta emitter, diagnosis and treatment of thyroid problems
Caesium-137: low-intensity gamma ray source for sterilisation
Carbon-14: radiocarbon dating in archaeology
Uranium-238, lead-206: geological dating
Nickel-63: light detectors in cameras
Gold-198: diagnosis of liver disease
Iridium-192: gamma emitter, cancer therapy implants
Stronium-90: beta emitter, electricity generating in remote places
Thallium-201: testing for coronary artery disease

6.28 CHAIN REACTION

An explosion occurred when a lump of metal about the size of a grapefruit became unstable and started a chain reaction. The amount of energy released produced an estimated blast equivalent to 16 000 tonnes of TNT and the World was introduced to a new and terrifying weapon. The uranium bomb exploded at Hiroshima on 6th August 1945. About 80 000 people were killed by the blast and another 70 000 were injured, many of whom were to die later. Three days later there was a similar explosion at Nagasaki, which brought the total death toll to at least 129 000 people.

Nuclear warfare had arrived. Its main destructive power was the enormous amount of energy released in the blast. But this was not all, for in the explosion there were radioactive isotopes in the dust produced. Much of the dust was airborne, enabling it to spread and slowly precipitate its radioactive poisons around the globe. In addition to the huge number of poisons from plants, animals, minerals and chemicals, we now had poisons that release radiation to kill. Death by stealth had become even more sinister.

With the interest in nuclear weapons, there came the need to test them. Atmospheric testing of nuclear weapons involved exploding the device on a remote island, or dropping it from an aircraft or from a tower on the ground, or with the device buried near to the surface. These result in large amounts of fallout containing radioactive poisons. Over 500 atmospheric nuclear explosions have been carried out, with a total destructive power equivalent to about 550 million tonnes of TNT.

6.29 FURTHER FALLOUT FEARS

Windscale nuclear facility in Cumbria, Britain, was set up to make plutonium for nuclear weapons and comprised two nuclear reactors. Uranium, contained in aluminium cartridges, was being fed into one reactor on 8th October 1957, when a problem occurred in the cooling system. This developed into a fire when hot uranium came into contact with the air. The fire burned for 16 hours during which radioactive iodine (iodine-131) was released into the atmosphere.

The fear was that the iodine contamination of the surrounding agricultural areas would find its way into the milk from cows feeding on the affected grass. Anyone drinking the milk would ingest radioactive iodine, which would concentrate in the thyroid gland, which takes up iodine when it produces thyroxine, a hormone containing a large proportion of iodine. As a result of this, the distribution of milk from over 200 square miles of farmland was banned.

When radioactive iodine is taken up by the body, the cells in the surrounding tissue suffer radiation poisoning due to the release of beta particles. Damage to the DNA in the cell goes on to produce mutations that can lead to cancer.

In the Windscale disaster, there was a problem with a control valve which led to the incident. In all new technologies, however well researched, planned and managed there are always going to be outcomes that could not have been predicted. It is all to do with the learning curve. There may be claims that a thorough risk assessment has been carried out, and it shows the procedures to be perfectly safe, but such words are wishful thinking. Or do the people that write this salesmen spiel really believe they know everything?

The learning experience continued as other nuclear reactors went wrong with disastrous consequences. Nuclear fuels and radioactive waste ended up in the environment and spread poison Worldwide. Three Mile Island, Chernobyl and Fukushima come to mind. There are still aspects about nuclear technology that are new and unclear.

6.30 FROM FOSSIL TO FISSILE

Our energy-thirsty industrial World needs electricity generation to increase. This must be done year on year as we use more electricity to enable the high standards of living and the growth of electronic devices. Furthermore, the so-called 'under-developed' countries are opting for an industrial future, which adds to energy demands.

To meet these needs using fuels based upon fossil carbon will surely make matters worse. In order to reduce carbon emissions from electricity generation a major step forward would be to switch from fossil carbon fuels to nuclear fuels. Comparing the carbon dioxide emissions for some primary source of energy, puts nuclear energy ahead of the rest. The carbon dioxide releases, per kWh of electricity produced are estimated as: nuclear fuel 12 g of CO_2; coal 820 g of CO_2; gas 490 g of CO_2.

Some would argue that uranium in a fission reactor produces no carbon dioxide emissions. However, the figures refer to a life cycle analysis in which the construction and decommissioning of the nuclear power plant is taken account of. This involves, for example, the manufacture of cement for the huge amounts of concrete required to contain the reactor. The extraction of the metal and the machining of it to make the turbines and generators will be about the same as for a fossil fuel power station.

Although there are somewhere around 500 nuclear reactors in operation at present, the technology is not yet complete, especially with regard to the containment of treacherously poisonous waste that is produced in the fission reaction. Most of the chemical waste that industry produces will, given time, either bio-degrade or chemically degrade into simple substances that then join their respective cycles (water, carbon, nitrogen and others) and become a part of nature once more. The waste from nuclear reactions is a challenge. We can't neutralise it, or incinerate it to harmless chemicals, because the products of nuclear reactions are elements; they are already broken down, but they are radioactive.

6.31 DIRTY BOMB

There are conventional bombs that use chemicals to produce a high-energy blast capable of killing people and destroying property. These bombs typically use TNT as the explosive, which means the products from the explosive reaction are mainly carbon dioxide and water vapour, which are not poisonous. The weapons effect comes from the blast.

Nuclear bombs, produce the energy for the destructive blast from the nuclear reaction occurring when uranium metal reaches its critical mass and starts a chain reaction (Figure 6.1). Vast amounts of energy, that far exceed the energy from chemical explosives, are released. The blast causes massive damage, but this is not all. A nuclear bomb is a double-edged sword. In addition to the damage caused by the explosion there is the

Figure 6.1 A nuclear bomb obtains most of its destructive power from the blast, but it also kills due to two types of poisoning. There is the radiation poisoning experienced by people absorbing the intense radiation, and there is the toxic effect caused by the radioactive dust produced. Radiation poisoning affects only those near to the core of the explosion. Radioactive dust also affects them some time later as fallout occurs. People at greater distances may also be poisoned as the radioactive dust spreads through the atmosphere. © Shutterstock.

damage to people from the radiation, and from radioactive particles the air. These eventually settle to produce radioactive fallout and this may be hundreds of miles from the centre of explosion. Dangerous pollution follows as the particles are dispersed in water and soil.

A 'dirty bomb' (or radiological dispersal device, RDD) is a terrorist device that uses conventional chemical explosives to create an explosion to blast out poisonous substances. These are often radioactive waste or radioactive materials from redundant or lost equipment that contained a radioactive source. Thus, the lethal action of a dirty bomb is from radioactive materials, which poison the human body through the ionising radiation that is released. Radiation poisoning will produce casualties and death and render the area unfit for habitation.

If uninhabited, the area becomes a no-go area and so the enemy is denied access. There is also the fear factor, which makes the dirty bomb a weapon of mass panic rather than a weapon of mass destruction. Although the area may become a no-go area, the buildings and infrastructure will be left intact and available for use at a later date, perhaps after a clean-up operation like those carried out after accidental nuclear explosions, such as occurred at Chernobyl.

6.32 ASSASSINATION

Nikolai Khokhlov was a Russian security official who defected to America and revealed details of KGB activities. In 1957 there was a failed attempt upon his life. He survived after treatment for thallium poisoning. This failed assassination is thought to be the first time a radioactive isotope was used to poison someone.

There are two distinct mechanisms by which thallium can cause poisoning. The first is through its chemical reaction in the body. Here, thallium mimics the body's normal potassium, which enables it to pass through cell membranes and damage the proper functioning of the cell. The second way in which thallium causes poisoning is by means of the radiation that is emitted by the thallium-201 isotope. In the body, this radiation damages surrounding cells to produce the symptoms of poisoning.

In London in 2006, Alexander Litvinenko, a Russian dissident, was poisoned with polonium-210. This element emits intense alpha particle radiation. When it is inside the human body, the cells of his liver, kidneys and bone marrow are constantly irradiated and ultimately destroyed. Litvinenko died of multiple organ failure.

6.33 VILE VOLATILES

Like the mad hatters, there were many people poisoned in the course of their work through inhalation of poisonous gases, vapours and dusts. Today, in the developed World, this is less of a problem, but still exists and, in the newly industrialising countries many workers are poisoned through exposure to chemicals. Where toxic substances are used for industrial purposes there will always be a risk, and this risk is often made worse by inadequate ventilation.

In the safety and comfort of our homes, we would not expect to be exposed to toxic chemicals. However, in the modern house, which is draught-proof and warm, there is a plethora of synthetic materials, and many of these are releasing vapours that we inhale. Electrical equipment that warms up to relatively high temperatures is a problem. Many of the gases and vapours are collectively known as 'volatile organic compounds' (VOCs). The quality of indoor air may not be as good as we assume.

Recently painted surfaces and new furnishings containing synthetic textiles and other man-made materials can release unhealthy amounts of VOCs. People who live in cold climates, keep the windows closed, have the central heating permanently on and spend most of their time indoors, are especially at risk. Newly built houses can have poor indoor air quality (IAQ) from the out-gassing of building materials.

Most of the risks relate to long-term exposure and are therefore chronic poisoning. But, acute poisoning can occur when the occupants inhale a single dose of carbon monoxide released into the air from badly maintained appliances, such as central heating boilers and water heaters that use natural gas. When an appliance is in poor condition it may not carry out complete combustion of the gas and, instead of releasing carbon dioxide, it gives out carbon monoxide.

Fortunately, we no longer have to worry about being poisoned by the gas itself. Some years ago the municipal gas supply piped into homes, contained carbon monoxide making it poisonous. Fortunately, this gas had a strong smell and any leakage was quickly detected. During burning of the gas the carbon monoxide turned into harmless carbon dioxide.

Today, gas for domestic use is natural gas, which is free from toxic chemicals, or almost. Natural gas has no odour and leakages could go by undetected so it is given a distinct smell by adding a stenching agent such as tetrahydrothiophene. When a leak occurs, the foul smell is soon detected. As all gas pipes and fittings leak, hopefully to a negligible extent, then gas enters the air. Minute amounts of this stenching agent will always be in the indoor air. However, when the gas is burnt the stenching agent is turned into sulphur dioxide along with the combustion products of the gas itself.

The IAQ, in the office and in the home, is compromised by the presence of chemical vapours. The main worry is from VOCs evaporating from synthetic materials, and these include acrolein, formaldehyde, benzene, vinyl chloride and others. And then there are the natural substances given off by the mites that live in the carpet and release chemicals such as geosmin, thujopsene and borneol. Houseplants are effective at keeping down the VOCs such as benzene as they constantly absorb these.

Ozone is produced where there is strong ultraviolet light, such as in a photocopier. In modern jetliners that fly for long periods high in the atmosphere the ozone can be a problem. Ozone is naturally produced from atmospheric oxygen when the intense ultraviolet at high altitudes turns the oxygen into ozone. These aircraft must take in fresh air from outside for the cabin air. The incoming ozone reacts with VOCs inside the cabin and produces irritant chemicals, which are inhaled. However, many aircraft are fitted with ozone filters, through which the outside air is passed before entering the cabin.

Is it any wonder that living the synthetic lifestyle is associated with so many cancers? At home we might consider opening the windows more frequently rather than expecting the unwelcome volatiles to disappear through leakage of the door and window seal.

In the days when aerosols were all the rage, the problem of CFC (chlorofluorocarbon) poisoning of the atmosphere was discovered. These chemicals were used to generate the pressure inside the aerosol can to produce the spray. The manufacturers of the aerosols thought that the CFCs just vanished into the atmosphere. What was happening was that the CFCs were building up in the stratosphere and causing a sequence of chemical reactions that created a hole in the ozone layer. The result was an increase in skin cancer caused by an increase in ultraviolet light.

6.34 WARMING OR DIMMING

Burning a litre of petrol produces a litre of water when the water vapour from the exhaust condenses. In effect we are taking fossil hydrogen from underground and turning it into water. Might this also be a factor in the increase in levels of rainfall? With regard to our gaseous effluents it is carbon dioxide that gets the Lion's share of attention, with methane coming second. Global warming is an observed fact and sea levels are starting to rise. Another example of the saying 'What goes around comes around'. We now also have global dimming due to increased levels of dust going into the atmosphere from transport, coal burning, metal smelting, cement making and other man-made activities. It is suggested that the cooling effect due to global dimming will offset global warming. Or is this wishful thinking?

6.35 TRIAL, ERROR AND IMPROVEMENT

Mankind's progress from the Stone Age to the Space Age is explained by his ability to think creatively. In this respect he is clearly distinct from animals. In general, animals spend all of their waking hours searching for food, defending themselves, and finding a mate. But humans, once they had learnt to live in communities with their animals in fields, and crops close to hand, had time to spare.

The archaeology shows us that, later on in the Stone Age, mankind was making pottery. This was the first example of humans making a material that does not exist in nature. It represents the beginnings of trial and error. Experiments must have been performed to move from a lump of clay baked hard in the embers of a charcoal fire, to shaped pots capable of holding liquids and with the right properties to enable their use as cooking pots.

There must have been a good deal of 'what if' experiments. In order to make such pots people had to learn about the types of clay, whether or not to mix in other materials such as grit, how to form it into the required shape, how long to allow it to dry, how hot to fire it, and the length of time for it to be fired. When an experiment failed to give the expected outcome they would take it as a learning experience and start again – back to the drawing board, as we would say today.

CHAPTER 7

From Poison to Prison

Examination of dead bodies is carried out to establish the circumstances of death. In some cases, the body will be that of a recently deceased, person but there are occasions when the body is from someone who died months or years ago. A human corpse is totally bio-degradable and the processes by which it degrades begin within hours of death. The degradation eventually results in the corpse being broken down into simple compounds such as carbon dioxide, water, ammonia and hydrogen sulphide. These enter the environment and then become available to make new living systems in nature's recycling scheme.

The body of someone who died from poisoning only an hour ago, clothed and indoors will be in relatively good condition and still warm. In contrast to this, a body that has been left naked and concealed in the undergrowth for a few weeks is going to be so badly decomposed that most identifying features have gone. Despite this, if the person dies of poisoning, then the poison may still be present. During the time the body was concealed it may also have suffered in that it provided an easy meal for a variety of living creatures; a human body is simply a large piece of meat so far as the fox is concerned.

In many instances it will be apparent what the cause of death is, but the body must still be examined to provide the proof. If poisoning is suspected, there will be a need to examine the

Poisons and Poisonings: Death by Stealth
By Tony Hargreaves

Published by the Royal Society of Chemistry, www.rsc.org

inside of the corpse. A living body works according to well-regulated chemistry, but the chemistry inside a dead body is chaotic and influenced by the surroundings in which the body is found.

Any poison present in the body may change due to the influence of the body's decomposition reactions and the chemistry of the environment around the body. The poison may break down into simple substances that are indistinguishable from the body's normal decomposition products; it may concentrate in a particular part of the body; it may alter the course of putrefaction; it may leach out and pass into the soil; it may accumulate in the flesh of maggots that are feeding upon the dead body.

All of these must be considered in the search for material evidence. Where there is no remaining body, such as in a cremation, there is a tendency to think that any poison present will have been destroyed. This is certainly the case for poisons based upon organic compounds, but certain poisonous metals can survive the cremation process and be detected in the ashes by modern analysis instruments.

7.1 BODY CHANGES

A fresh body carries a lot of valuable evidence, but it takes little time for the body to change with subsequent loss of evidence. In efforts to understand the death, the investigator needs to establish the time of death, the identity of corpse, and cause of death. Time of death often takes priority, as this may link with other events that are essential pieces in the jigsaw. Clearly, a fresh body must be examined as a matter of urgency. If delayed, there is loss of evidence. Estimates as to the time of death become less and less accurate as the time elapsed since the death occurred, the 'post-mortem interval', increases.

In addition to concerns over loss of evidence, the matter of contamination from the investigators has to be considered. It is all too easy to contaminate the area with a trace of material that can send the investigation off in a false direction. Locard's Principle applies just as much to the investigator and crime scene as to the culprit and victim. The trace left may not be visible to the naked eye, and may even need powerful analytical methods to detect and identify, but, all the same, it can be vital evidence.

In general, the body is not touched or moved until its immediate surroundings have been examined, photos taken, and notes made. This assumes, of course, that the victim has been declared dead. The greatest of care is practised by the forensic scientists to minimise the loss of evidence. However, as soon as someone enters the incident scene, it is inevitable that some of the evidence is destroyed.

7.2 WHERE'S THE BODY?

A murder is suspected but there is no corpse. Is it concealed or have attempts been made to destroy it? Burial both conceals and, ultimately, destroys the body. Being the most common choice of disposal by the killer, we need to examine the process in detail. To move a dead body to the place where it is to be buried is often problematical. The corpse of an adult is not easy to handle and often has to be dragged. This explains why most bodies are found downhill from the murder scene. If the body is in the rigor mortis stage, this poses an additional problem in that it cannot be bent to fit a particular space such as a car boot or a regular shaped burial pit.

However, once the body is delivered to the burial site we might expect the process to be relatively easy: just dig a hole, shove the body in and cover with soil. But, as every gardener knows, to dig deep is no easy task. As soon as the spade has cleared the few inches' depth of topsoil, there is heavy clay and/or rocky subsoil. Removal of this typically requires the use of both pick and spade. Furthermore, it needs a lot of energy and it is time consuming – something that the killer does not want.

Faced with these problems, someone attempting to bury a body is forced to opt for a shallow grave. A major disadvantage of this is that the body, or parts of it, may surface due to the efforts of animals or someone doing a bit of weeding or raking over to tidy up the area. Burial of a body in a grave dug in the cellar is certainly going to be away from prying eyes, and will allow time enough for a thorough job to be done. However, it will not be long before the putrefaction gases permeate the house and surroundings. Disposing of the body in water such as the sea or a lake also brings its own set of problems.

A well-known example of shallow grave burial is to be found in the case of Christie at 10 Rillington Place just after World War II. While he was tidying up the back garden he unearthed a skull and a thigh-bone of one of his victims. Rather than dig a deeper grave he took the skull away and left it in a derelict house, where it was later found by children out playing. They took it to the police, but it was thought to be from a victim of the German air raids and was destroyed. The thigh-bone, however, was kept by Christie and used for propping up the listing garden-fence.

A killer, wise to the problems of burial in ordinary soil, may choose soft spongy peat that can offer a depth of several feet before the rocky subsoil is encountered. Burying a body in peat may not be quite as permanent as the murderer hoped. For example a body buried in peat, such as on the moors, may soon come to light. Unlike soil, peat is often in thick layers that take on fluid properties when saturated with water. In this state the peat is prone to shift, especially if it is on a slope. It also splits open into clumps from which bits are washed away in the next downpour.

In the evil perpetrated by Ian Brady and Myra Hindley in the 1960s, the bodies of murdered children were buried in moorland peat on top of the Pennines. When police investigations began, one officer came across a bone sticking out of the surface of the peat, and there can be little doubt that its appearance at the surface was in some part, at least, due to the fluid properties of peat.

7.3 TROWEL OR SHOVEL

Consider a scene where it is believed a body is buried and the process of its recovery then begins. The traditional approach was for the police to set about, with spades and shovels, and dig the earth away. Needless to say, a lot of valuable evidence could be lost in the process. These days, it is not simply a matter of digging up the body, but one of excavation of the burial. Those employed to do this are forensic archaeologists and they use the techniques of archaeological science. The trowel has replaced the spade. A typical start is to survey the ground that has been identified as a location in which there is a high probability of finding the grave.

A preliminary ground survey includes a simple visual inspection. Are there any bumps on an otherwise smooth surface? The burial of a body, involves putting a volume of extra material into the ground, and so there must be a similar volume of soil displaced. Visual inspection of the area may be extended to viewing the site from different angles and in different light. For example, when the sun is low in the sky the shadows cast by surface features can reveal a lot of detail that was simply not visible with the sun overhead. A covering of snow can also be beneficial in showing undulations in the surface. Aerial photography has also been used, especially for locating mass graves.

7.4 GRAVE FACTS

An adult of body mass 70 kg has a body volume of about 70 L. Burial in a shallow grave, without coffin, results in 70 L of soil being displaced. Piling this back on top of the burial leaves a distinct feature in the landscape.

If the grave is dug in an area covered with vegetation that had not had sufficient time to grow back, this would also be soon detected, as would footprints in the freshly disturbed soil. When the shallow grave has been there long enough for decomposition of the body to be underway, the volume of the body will decrease. This is caused by the soft tissue liquefying and draining away into the earth or forming putrefaction gases that diffuse out into the open air. For the 70 kg body, the water alone amounts to 50 L. The grave sinks and this leaves a noticeable depression in the surface. In some cases, the abdominal cavity of the body collapses first because this is entirely soft tissue and rapidly putrefies to liquid and gas.

This contrasts with the thoracic cavity, which is supported by the ribs and collapses more slowly. The depression in the soil over the grave may therefore reflect this. It is noteworthy that in the burial grounds of old churches, where many generations of locals were interred, the ground is raised above the natural level due to the volume of bones beneath it.

7.5 INSTRUMENT SURVEYS

Yet a further way of looking at the site involves infrared light (IR) rather than visible light. If the shallow grave is relatively recent, and the body still fresh, it may produce an image using IR detection equipment. IR is emitted by living humans due to their normal body heat from respiration. In a dead body there is no heating mechanism (from respiration), just as there is no cooling mechanism (from homeostasis), and the body becomes the same temperature as its environment.

Generally this involves the body cooling. Once the body is at the same temperature as its surroundings, its IR becomes just another part of the background IR. However, sometime later the body begins to decompose, and this involves exothermic chemical reactions that release heat. By this means, the soil above the grave warms slightly and emits IR light.

The ground survey methods include measuring the physical properties of the ground. Resistivity and magnetometry measurements are made by taking readings over a large area that has been marked out as a grid, typically on a square metre basis. Where there are buried features, not necessarily dead bodies or recently disturbed ground, the measurements differ from those of the surrounding ground that is in its natural condition.

Resistivity is a measure of how well the soil conducts electricity. High resistivity is low conductivity and *vice versa*. The amount of moisture in the soil has a profound effect upon resistivity. If something is placed in the ground that improves drainage, the soil above it will be drier and the resistivity will be high. If, on the other hand, an object that blocks drainage is placed in the ground, such as a sheet of polythene, then the soil will contain greater moisture than the surrounding soil and the resistivity will be low.

When a dead body is decaying in the ground, it produces high concentrations of ions such as ammonium, phosphate, nitrate, calcium and potassium. These dissolve into the soil moisture and reduce the resistivity. Even when a body has been reduced to a skeleton there is still a release of ions, as the bone demineralises and releases calcium and phosphate. The concentrations of ions from this source are too small to affect resistivity to any measurable extent, but the ions could be detected by chemical analysis such as liquid chromatography (for phosphate) and atomic absorption spectroscopy (for calcium). Plant growth is affected by the presence of decaying human remains in the ground. For example, the ions mentioned above are plant nutrients and will boost plant growth making the spot where the grave is stand out against the surrounding, possibly nutrient-starved vegetation.

Magnetometry measures the magnetism of the rocks and stones in the ground, especially when the rock contains iron minerals as in fact, many rocks do. If the rock has been undisturbed for a long time, its weak magnetic field will have aligned with that of the Earth's magnetic field. Where the rocks and stones are disturbed by digging, the magnetic alignment changes and it will take many years for it to re-align with the Earth's magnetic field. In this way, magnetometry can reveal recently dug ground that may otherwise not be seen with the eye or detected by the resistivity measurements.

Dogs are useful in sniffing out and locating a body. Probing the ground with a metal rod and then sniffing the section of the rod that has penetrated the ground is also an option. The tell-tale smell of a decomposing corpse is due to volatile compounds known as 'ptomaines'. These chemicals are bases, and therefore are neutralised by acids such as the humic acids found in peat. Once neutralised there is virtually no smell. Thus, the absence of the smell of these chemicals does not mean there is an absence of a body.

Sniffing the chemicals associated with a decomposing corpse is now being put onto a more scientific footing with the development of portable analytical instruments. The most common, which is as sensitive as a dog's nose, is a small gas chromatography instrument with a flame ionisation detector.

7.6 BURNT BODIES AND BONES

Burning a body by pouring an accelerant, such as petrol, onto it and setting it alight may cause sufficient damage to make the task of identifying it much more difficult. This delay in the process of detection then allows more time for the culprit to conveniently disappear.

However, such crude attempts do not dispose of the body as the culprit might have hoped for. To burn a human body is extremely difficult because the body is mostly water. A typical adult of around 70 kg body mass contains about 50 kg of water (same as 50 L of water). Before the body fats and proteins can start to burn, all that water must be boiled off, which requires much energy. Until the water is removed, the body cannot exceed 100 °C, the boiling point of water. Once the water has gone, further heating causes the temperature to increase and the body fats melt. If heating is continued, the fats reach their flashpoint and ignite – rather like the over-heated chip pan igniting. The burning fats are, in effect, the fuel. Although proteins also burn, their energy output is less than that of the fats.

The requirement for a large heat input to burn a body explains why it takes a municipal cremator about an hour and a half at temperatures up to 800 °C to render an adult body to gas and ash. The amount of mains gas that has to be burnt to achieve this is around 40 000 L. Clearly, a 5 L can of petrol is no match for this.

7.7 SPONTANEOUS COMBUSTION

At this stage it is appropriate to consider the bizarre idea of spontaneous combustion. This is a theory put forward to account for the presence of a burnt body, or part of a body, where there was no burning of the surroundings. The body has burnt in isolation to its immediate environment, which has been apparently unaffected. Can the human body self-ignite? It is a theory, but where is the evidence? From what we know about the difficulty in burning a corpse, we would be justified in doubting the notion of spontaneous combustion. But an open mind must be kept, for it is a fact that strange things do happen that we cannot explain.

The universe is a strange and wonderful place and humans know only a tiny fraction of its mysteries, despite the tendency of some to imply that we can explain everything that occurs on Earth. We may have heard accounts of spontaneous combustion, and some people may say this is evidence enough. It is often the case that someone describing an event or sequence of events reports their own version of it. And another person, seeing the same event, may give a different account. There is doubt. Who do we believe? This now places us in a position of believing rather than knowing.

What is really needed is physical evidence. What exactly would we look for? Our science can guide us. We know that for the body to burn it must, or a part of it must, dehydrate so that the fat can become sufficiently hot to self-ignite (auto-ignition temperature). The living body is at a temperature of 37 °C and when dead is normally less than this. Of course, if the individual dies and is left in a hot place, their body temperature will certainly increase. But unless that hot place has a temperature going on for 200 °C, the body fats will not ignite of their own volition.

There is here a slight twist. Can the human body produce its own heat and reach the auto-ignition temperature for fat? Certainly, a dead body can heat up due to the exothermic decomposition reactions of the internal organs. And if it is well insulated there is a measurable temperature increase, but it amounts to only a degree or two. But that is the general rule and there may be particular cases that are the exception to the rule. For example, a compost heap that contains a mass of dead vegetation and is sat outside in the cold can suddenly catch fire. Yes, compost heaps occasionally undergo spontaneous combustion. A coal heap, if it gets wet, can also catch fire entirely on its own: no external action such as lightning to raise its temperature is needed.

The compost heap is worth considering in some depth. The bulk of it is organic matter, water and air. This combination is ideal for bacteria to live, and as they do so they produce heat. As the heat increases it becomes too hot for the bacteria and they die, but by this stage chemical reactions have begun and these also produce heat. The reactions continue and the heat accumulates faster than it can escape, resulting in an increase in temperature. A point is reached at which ignition occurs and the compost heap proceeds to burn.

The human body has organic matter, water and air in contact and there are bacteria in and all around it. However, the temperature of the living body is regulated, and surplus heat readily removed. A dead body may present the right conditions for spontaneous combustion, but there is far too much water to be removed. Maybe if the body had partially dehydrated in a process such as mummification, it is possible that a sequence of reactions like those in the compost heap could operate.

It is worth considering the effect of certain poisons in relation to spontaneous combustion. If the person had died of alcohol poisoning, they would have distributed through their body an amount of ethanol, and ethanol is highly flammable. However, it is in aqueous solution and diluted with a large volume of water and far too weak to ignite. Other poisons may also be considered. Heavy metals are of interest here because they can behave as catalysts and speed up or alter the course of chemical reactions, or introduce new reactions. A reaction that is made to speed up, also gives out heat at an increased rate and has the potential to create 'hot spots' where the heat is building up faster than it is being removed.

Of the bodies examined that fit the idea of spontaneous combustion, forensic scientists have offered a sensible theory to explain it. This is the 'wick theory'. The idea is that part of the body was near to a source of heat such as an electric fire. As that part of the dead body heated up it dried out and became hot enough for the body fats in that particular area to melt. The body fat then being a liquid soaked into surrounding clothing. As the temperature increased further, the fat ignited and the fat-soaked clothing acted as a wick and supplied the flame with more fat.

In considering other possibilities we might focus upon electrical discharge, as having sufficient energy to dehydrate, melt and ignite body fat. The discharge may come from electrical equipment or it may be natural. Natural electrostatic events such as lightning strikes are worth examining. Ball lightning is especially interesting in this context.

Ball lightning is thought to be lightning that has broken away from a main strike in a cloud-to-ground discharge or a cloud-to-cloud discharge. The latter involves lightning inside clouds and may pass unnoticed by us. Eyewitness accounts of ball lightning are fairly common. Many of these accounts are offered in good faith and seem credible. On the other hand, there are some descriptions that are just too far-fetched. However, an open mind is important, for new phenomena are still being discovered. For example, two new types of lightning have recently been observed high in the atmosphere: the mysteries of red-sprites and blue jets.

7.8 DEATH AND DECAY

Death is not an instantaneous event, but is a series of changes that occur as the different types of cells gradually cease functioning. This is a general rule and there are exceptions. For example, death can occur in an instant during a nuclear explosion. Under such conditions the whole of the body, including bones, dental amalgam and implants, are turned into gas within milliseconds. However, in this book we are dealing with poisoning, which is far from instantaneous.

Different cells die at different rates, with the brain cells being the first to perish. Other types of cells continue to live even when the blood circulation has stopped and they are no longer provided with the materials for their respiration, fuel (glucose) and oxygen, for they have some reserves to tide them over. Sperm cells, for example, can live on and still be swimming around, looking for an egg to join company with, for around 36 hours after the blood circulation has stopped. However, brain cells cease functioning within minutes of their fuel and oxygen supply coming to a halt.

7.9 BRAIN DEATH

Once the blood stops flowing it takes about three minutes for all the brain cells to be irreversibly changed. It was chemistry that was responsible for their activity. One of the functions of brain cell activity was to produce consciousness. What happens to consciousness as death takes over is unknown, we can only guess.

Guessing, or any other form of conjecture, is subjective and this makes it a creative exercise. As such it cannot be allowed to be a part of the scientific process, a process that must remain totally objective and offer conclusions based upon evidence. Some would say that creative thinking has no part in science; it is an exercise of no value; it sometimes involves crazy ideas; it is belief-based. We might counter these comments by saying that creative thinking or brainstorming stimulates the mind to think laterally rather than vertically. The next few paragraphs explore some creative thinking in relation to the changes in the brain at the point of death. We might treat it as a science fiction interlude.

If the brain cells are the source of consciousness, then making changes in those cells may change the quality of consciousness. At the moment of death we can reasonably expect some change in consciousness, at least for a short amount of time. Of course, in the longer term, the brain cells face total destruction as they decompose into simple compounds such as carbon dioxide and water. There can be no consciousness then. Thus, we can go on to say that the dying person (not the dead person) experiences a changed reality. This is a transition state in which the thought processes change in a few fleeting moments. An additional effect is that the input that the brain receives from the sensory organs is slowly shut off.

Another idea to pursue is that the dying person's time-base changes, and what to us, in our real time, is three minutes, to the dying person could be experienced as a whole lifetime in itself. We know that our time perception can change. For example, if we are involved in a serious accident we may experience the slowing down of our time-base. And there are drugs such as speed that are said to have a similar effect.

It is said that in the initial stages of death, the deceased can still hear, suggesting that, of all the senses, hearing is the last one to stop. So, don't speak ill of the dead for they may be able to hear you. The other senses, it would seem, cease to function fairly quickly. But what are these other senses? The obvious answer is that they are: sight; touch; taste; and smell. Is this the whole truth or are there other senses? It is now a recognised fact that migrating birds have an extra sense, they can detect and follow the Earth's magnetism. And there are those who suggest that humans can unknowingly send and receive signals telepathically between family and loved ones.

Could this be electromagnetic radiation just like radio/TV/IR signals? If this is the case, the possibility of the consciousness being transmitted to others at the point of death deserves consideration. The crucial thing here is that this would enable consciousness or some elements of it to continue. Certainly, the atoms and molecules that make up the human body will, after decay of the body, go on to make up other human beings. For example, every one of us contains water molecules that once were a part of people and animals in the past. Matter can be neither created nor destroyed by chemical reactions. Energy can be neither created nor destroyed by chemical reactions, but can change from one form to another.

In the process of dying what happens to consciousness? Does it simply dissolve away or does it go elsewhere? If consciousness is an energy phenomenon then it cannot be destroyed. At the point of brain death, is the energy transmitted as electromagnetic radiation and is it received by a living person? We can only ponder. However, the newly developing science of neurochemistry may, in the years to come, provide us with the clues to enable objective comments.

It is worthwhile here considering the endorphin chemistry of the brain and how this might change during the dying process. Pain certainly plays a part in many deaths, probably all. For our own death we can only hope that it is rapid and that we do not suffer days of agony, a feature of many deaths, especially poisonings.

The brain responds to pain by releasing compounds known as 'endorphins' (endogenous opiates). These compounds are the ones that, in addition to relieving pain, give us the feel-good factor and are similar to the morphine-related drugs sought by many an addict. It is possible that there is a release of endorphins during the onset of death. If this is so, then dying may be the best feeling you've ever had. In considering consciousness at the point of death we are taken into the area of different belief systems in which, for most, the idea of an afterlife plays a central role.

This goes on to ideas of the supernatural. We often hear of people who promote themselves as having special powers. On learning of the search for a body they offer their services, claiming to be able to locate it. Sometimes their interest is helpful, other times it is a hindrance. Should we ever listen to these people? Among scientists, who should have open minds, there is a tendency to instantly dismiss such ideas. But it does leave many people wondering.

In considering what happens to the mind at the moment of death we might wish to examine the ideas of the Tibetan Buddhists and their 'bardo state', something discussed in *The Tibetan Book of Living and Dying.*

7.10 CORPSE CHEMISTRY

Chemistry is what enables the human body to function, and poisons disrupt those reactions and stop the body working. The body breaks down due to decomposition chemistry and the forensic scientists use analytical chemistry to study all of this.

Clearly, if we are to understand poisoning, then a study of chemistry is essential. In the following paragraphs we take a look at the chemistry of death and decay. The changes may be affected by the environment in which the corpse is left, along with other factors: whether the body was clothed or naked; what the state of health was prior to death; and if attempts have been made to destroy the body, such as efforts to burn it. The processes are outlined below, but the order in which they occur can be affected by the circumstances of the death and what has happened to the body immediately after death – the 'postmortem conditions'.

Soon after the blood has stopped flowing to the cell's respiration, the chemical reaction that produces heat energy, ceases. No more heat is produced and, where the ambient temperature is lower than that of the body, as is normally the case, the body cools. Making measurements of body temperature can fairly accurately establish the time of death if they are made soon enough. The core body temperature is measured by placing a thermometer in the rectum or into the liver.

Consider some typical figures. At the time of death the body is 37 °C, assuming the person was in normal health and not running a fever. If the ambient temperature is about 18 °C, after 10 hours the core temperature will be 28 °C. As the body cools the rate of the chemical reactions inside it decreases, and this may help preserve some organic poisons that are easily broken down. Naturally, this cooling will be affected by ambient temperature and other factors such as clothing.

7.11 DRY EYES

The eyes soon show changes. The protein that forms the cornea contains water bonded to the protein molecules, and is known as 'water of crystallisation'. In the living, the outer surface is kept moist by the process of blinking, which wipes the fluid over the cornea's surface. Thus, a balance exists between the water in the protein structure and the thin film of water on the surface. At death, the blink mechanism ceases, the eyes remain open and the water on the surface evaporates. This draws water out of the crystalline protein, resulting in the protein dehydrating, becoming opaque and giving the eyes a cloudy appearance.

Another effect taking place in the eyes occurs when the cells in the retina begin to break down. During this process potassium ions are released into the optic fluid. In some instances, the optic fluid is sampled by the forensic pathologist and sent to the laboratory where the concentration of potassium is determined. The result helps in assessing the post-mortem interval. To obtain the sample, the needle of a hypodermic syringe is pushed into the eyeball and a portion of the fluid drawn out.

7.12 LIVOR MORTIS TO RIGOR MORTIS

This is also known as 'hypostasis' or 'post-mortem lividity'. Once the blood stops circulating, gravity takes effect and the blood sinks to the lower parts of the body. The blood concentrating in these parts gives the appearance of a bruise due to the dark purplish blue colour. The first signs appear at about an hour after death and continue to develop for up to four hours.

Livor mortis does not occur in parts of the body that are under pressure, such as where it presses onto a hard surface or where there is tight clothing, because in these areas the blood vessels are compressed so that blood cannot flow into them. Similarly, if in the early stages of livor mortis, pressure is applied to the coloured area the colour disappears as the blood is squeezed out of the underlying blood vessels. The development of livor mortis ceases when the blood coagulates and loses it fluid properties.

Consider death by hanging. Soon after death occurs the blood sinks to the lower parts of the arms and legs. If, after an hour into the post-mortem period, the rope was cut and the body placed flat on the ground, the lividity would move to the lower parts formed by the new position. However, if the rope was cut many hours after death, then placing the body on the ground would have no effect upon the lividity, which would remain in the lower arms and legs. Thus, the condition can be used to indicate if a body had been moved, but to use livor mortis as an indicator of time of death is unreliable.

Where a poison is present in the blood, then high concentrations of that poison can be expected in the blood-rich areas, that is, where lividity has developed. For example, if the person died of carbon monoxide poisoning, there would be a high concentration of the poison where there is lividity. With carbon monoxide poisoning the colour is a cherry-pink rather than the purplish blue.

Rigor mortis generally occurs between 3 and 36 hours after death, and is seen as a stiffening of the muscles of the body. The time taken for rigor mortis may depend upon the amount of exertion the person had undergone just before death. Vigorous exercise immediately before death results in a fairly rapid onset of rigor mortis. In the extreme case of strychnine poisoning, the body is so exhausted due to the convulsions that rigor occurs straight after death.

What we experience as muscular cramp is rather similar. It occurs due to the build up of lactic acid in the muscles. The muscles tend to freeze due to a shortage of oxygen reaching them. In the case of cramp, the effect is temporary because the circulating blood will catch up on the oxygen supply, returning the muscle to normal. In a corpse, the effect is rather more long-lasting. In fact, the body would remain in a state of rigor mortis if it were not for subsequent changes such as the next one.

7.13 PROTEIN PUTREFACTION

This starts with a process known as 'autolysis', which occurs inside the cell, where the cell's own enzymes attack the protein that forms the cell membrane. The large molecules of protein break down into small amino acid molecules, resulting in the membrane losing its structural integrity and rupturing. The cellular fluid is released into surrounding tissue, with the overall effect that the tissue turns from a firm well-defined mass into a mobile fluid.

Migration of the cellular fluid containing the enzymes spreads the destructive action. At this point, the skin starts to break loose from its underlying tissue. Following the onset of autolysis the main stage of decomposition, known as 'putrefaction', follows. Putrefaction continues the breakdown of essential compounds such as proteins and fats. Where the proteins were those forming muscle tissue, this loses its strength and turns to a gel. Muscles that were rigid due to rigor mortis then become flaccid, and a stiff body becomes floppy.

Putrefaction is largely a bacterial process, the bacteria coming from the digestive system as its organs begin to break down. The intestines contain a lot of bacteria, and play an important role in putrefaction once the intestinal wall bursts open and floods the surrounding tissue with them.

As the main process relies upon bacteria, we can explain why certain poisons can delay the onset of putrefaction and, where they persist, reduce the subsequent rate of putrefaction. This has a preserving action upon the body. For example, when Napoleon's remains were exhumed for reburial in France they were seen to be in a good state of preservation, a sure sign of a poison like arsenic. It is worth noting here that poisons such as arsenic and other elements do not decompose, whereas poisons based upon organic compounds generally do decompose, alongside the organic compounds that make up the human body.

The arsenic may diffuse throughout the dead remains and even take on different chemical forms. For instance, it may be present as a mixture of arsenite ions, arsine gas and cacodyl. In the intestines it forms yellow arsenic sulphide due to its reacting with the sulphur in the proteins of the tissue it is attacking. The bright yellow deposit of arsenic sulphide is easily seen.

Bodies sealed in airtight lead sarcophagi can remain for centuries in remarkably good condition, due to lead acting as a preservative. Rather like arsenic, the lead kills the bacteria that cause putrefaction. Lead coffins, usually for important members of the community, are no longer used, and would nowadays be banned for environmental reasons, the main one being that lead ultimately ends up in the groundwater, and much of our potable water supply is abstracted from groundwater.

It is often said that bodies in our modern times putrefy slower than they did in the past. The explanation offered is that these days we eat a large proportion of industrial food that contains chemical preservatives, needed to keep the food in good condition, so it can be kept for a few months before it is eaten. There is a natural tendency to think that a corpse is completely static – as still as the grave. This is not so. Leaving aside the activities of burrowing maggots and slithering worms, the corpse is 'alive' with internal movements: gas bubbles form, rise and break; concentration gradients build up and cause fluids to move and churn; temperature gradients produce convection; diffusion of ions and molecules in fluids takes place; and gas cavities form and collapse. There is much activity inside the decaying body.

Putrefaction is a sequence of exothermic chemical reactions in which the amino acid molecules, from the decomposition of proteins, are broken down into smaller and simpler molecules. Large amounts of gases such as hydrogen sulphide and carbon dioxide are produced, along with the ptomaines, which are some of the worst smelling substances. Despite their retch-inducing stench, they are harmless but they tell us to keep away because the bacteria that produce them are not harmless.

The putrefaction gases build up and cause pressure that results in the abdomen bloating. The effect of all this is that the corpse starts to get warm, moves a little and makes rude noises as the high pressure of the gases is released. The increasing pressure in the intestines causes them to rupture and their contents spill out. Putrefaction continues and spreads, resulting in the internal organs turning to liquid and gas. As these are fluids they leak out of the body into the environment and the remains of the corpse takes on a flattened appearance.

During putrefaction, the chemistry is quite aggressive and capable of destroying many poisons, in particular the organic compounds that had so far survived. However, there are some poisons that will survive all of this. Heavy metals and arsenic, for instance, are elements and cannot be broken down to anything simpler, and they bond strongly with the relatively decay-resistant proteins of hair and nails.

7.14 CREEPY CRAWLIES FROM COFFINS

It is not long after death, sometimes less than an hour, when flies descend upon the body. As autolysis and putrefaction proceed, the large amounts of putrid-smelling gas broadcast the message further afield. Here is a supply of food. This adds to the number of flies and attracts other creatures out for a free meal. The flies are looking for food, somewhere to lay their eggs, and a supply of nutrients for their offspring. Preferred areas for egg laying are the exposed moist parts. For example a naked corpse has the following available: eyes; mouth; nose; penis; vagina; and anus. The nose is especially attractive because it is a cavity and offers an element of protection for the eggs and the emergent maggots (larvae).

There may also be cuts or other wounds that provide moist open flesh. Of course if the body is in a mutilated state there are opportunities galore. Birds and animals may also visit the body and eat, or take away, soft and easily accessible parts such as the eyes. Loss of part of the body by this means represents loss of evidence – much of the poison may have been concentrated in the part removed by the hungry visitors. However, the maggots stay until they have pupated, undergone metamorphosis to a fly, spread their wings and made off. During the time they are residents on (or in) the body, they are living evidence that can prove of great value to the forensic entomologist.

In the examination of a person suspected of being poisoned or having accidentally died of a drug overdose, a sample of maggots may be collected and taken for chemical analysis. If they have fed upon a corpse that contains poison, this will accumulate in their flesh. In effect they act as devices for concentrating the drug or poison. Identifying the type of maggot can provide crucial information. One important fact is that certain maggots are only to be found in a particular geographical location. For their egg-laying activities, some species of fly prefer woodland, others like an indoor environment.

This fact may be useful in establishing whether or not a body has been moved from one location to another. If the body is buried we may think it is out of the reach of flies but one species, known as the 'coffin fly', lays its eggs in the surface of the soil covering the body. The maggots that crawl out of the eggs burrow down through the soil to the corpse. Despite their name, the recipient body does not have to be housed in a coffin.

Other evidence comes from the changes the maggots undergo as they grow. For example, an estimate of the post-mortem interval may be obtained by studying the maggots' cast-off skins and pupal cases. It is known how long each stage in a maggot's development takes at, say, a particular temperature. Temperature is important because insects are cold-blooded, and their rate of growth therefore depends on how much heat they can absorb from the environment. Simply, in the cold they develop slowly, but warm conditions make them grow quickly.

7.15 METALLIC IMPLANTS

The final stage for the breakdown of the bulk of the soft tissue is carried out by fungi in a kind of fermentation process, that turns the last residues of fats into fatty acids, such as butanoic acid, which evaporate from the remains creating a cheesy smell. On completion of the fermentation processes, the soft tissue has gone and all that remains are the bones, teeth and hair. If in moist conditions, particularly wet soil, the protein collagen and the mineral calcium phosphate that compose the bones, will slowly start to break down. The collagen will be destroyed by the fungi and the calcium phosphate will dissolve as calcium ions and phosphate ions, seeping into the soil to be available for plant roots to absorb. Although bone will ultimately break down, if it is burnt bone it will survive for much longer.

With a modern day body we can also expect to find other materials that break down extremely slowly and last well into the skeletal stage. For example: dentures (acrylic polymers); a contraceptive coil (copper and polymer); prosthetic limb parts (titanium); and silicone breast implants (siloxane polymers). Dental amalgam, used for filling cavities in teeth, dissolves slowly in contact with soil, releasing its component metals: mercury; silver; tin; and copper.

Hair, made of the protein keratin, is surprisingly resistant to decomposition and unearthed corpses, hundreds of years old are occasionally found in which the hair has retained some features of the deceased's last hairdo.

7.16 PEAT MARSH

Up to this stage, most of the decomposition chemistry has come from within the corpse, with the environment, such as the soil, playing a limited role. An interesting example is the way in which acidic conditions affect the corpse. Wet peat, which has a low pH, due to its content of humic acids, can preserve a body. An example is that of the bog body, Lindow man, who forms a part of the British Museum collection. The body was discovered in 1984 by workmen cutting out peat at Lindow Moss near Wilmslow in Cheshire. Apparently he had died some 2000 years ago as a result of an Iron Age human sacrifice. Examination of his remains revealed that just before being killed he had eaten a meal of unleavened bread made from wheat and barley. His skull had been smashed and his throat cut before he was placed in the peat bog face down.

The remarkable state of preservation was due the humic acids preserving the protein in his skin by a chemical process akin to the tanning of leather. However, despite the peat acids acting as a preservative for the proteins, they have the opposite effect on bone. The acids demineralise bone by breaking down the calcium phosphate and removing the calcium and phosphate. In the body of a poisoned person, we can now appreciate that, under appropriate conditions, the poison may well be preserved along with the organic parts of the body.

7.17 HIDE AND SEEK

In most murders it is glaringly obvious what type of foul play was committed, simply by taking a look at the body. The evidence is there for all to see and the killer has to deal with the corpse as a matter of urgency, that is, if he has not made off. In poisoning cases, obvious signs of murder are absent and death would seem to have been due to natural causes, especially if the person had suffered from some serious illness. This is because poisoning causes damage inside the body where it cannot be seen.

Thus, the poisoner may simply rely upon the body being disposed of in the usual - legal - manner. Without doubt, many poisonings go by unnoticed. However, the killer should not be too confident at having got away with it. It does sometimes happen that evidence from another direction comes to light after the disposal of the body. The body may be exhumed, tests performed and residues of poison discovered. If cremated, the risk of discovering the poison is extremely small, however, we cannot say there is no risk because certain poisons, especially heavy metals that accumulate in the bones, can be detected in the ashes. A recent example of exhumation was that of Mrs K Grundy's remains. Traces of morphine were found and provided crucial evidence that helped convict Dr Harold Shipman. Poison remaining in a buried body may break down or migrate away from the body, depending upon the conditions.

7.18 GRAVE WAX

Burial often involves contact with water, and this can have a profound effect. For example a body dumped into water and weighted to prevent flotation may develop a waxy coating known as 'grave wax' (adipocere) and this helps slow down decomposition and restricts the leaching out of a poison into the surrounding water.

The effect of lead in retarding the putrefaction process has already been noted. Returning to the lead coffin, we find that bodies from Medieval times have been found in surprisingly good condition. After placing the body inside the coffin, the lead was soldered to prevent air getting in and nasty smelling gases getting out. Inside would be sufficient oxygen for the putrefaction reactions to begin, but this would soon become exhausted and decomposition would cease.

7.19 DEHYDRATING THE DEAD

Mummification of a body occurs when the body dries out. If this dehydration process is significantly faster than the decomposition, then the body will mummify and much of the decomposition will be stopped. The process of dehydration is similar to preserving food by desiccation. Bacteria that cause a body (and food) to decompose can work only in an aqueous environment. Take away the water and the bacterial decomposition stops. This process helps preserve many poisons because the reactions that generally destroy the more sensitive organic poisons are water-based reactions. Bodies found in air cavities in buildings have shown a good state of preservation due to mummification.

Other means of body disposal/preservation, whether legal or clandestine, exist, and are interesting but of little importance to the forensic scientist. To name but a few: smoking; pickling; freezing; bog burial; cooking; and eating. The latter brings to mind the activities of Sweeney Todd (real or fictional?) who, with the assistance of his partner in crime Mrs Lovett, had his victim's flesh made into pies and sold in her shop. It may seem incredible, but some killers have attempted to dispose of their victims by cooking the body and eating the flesh. Of course, if the body of the victim is the result of poisoning this would be highly risky.

7.20 BODY FARM

Aimed at understanding what happens to a dead body when it is left in the open for nature to take its course is the so-called Body

Farm. The Anthropology Research Facility is a part of the Forensic Anthropology Center (FAC) at the University of Tennessee, and was established in 1980. The Facility was formerly a hospital waste area next to the medical center. It was prepared and the first donated body arrived in 1981. The body was studied and the rates of various changes recorded. A priority in studying dead bodies was to obtain data that would enable anthropologists and pathologists to give a more precise estimate of the time of death.

Dramatic changes took place over the first two weeks of the body's exposure. The hair had slid off the head as a mat and was stuck to the ground by greasy slime in a pond of body fat. The fact that shortly after death the skin becomes loose and falls off would account for this. The abdomen had originally bloated but then collapsed and clung to the ribcage.

In the heat of the Tennessee summer, a body is typically converted to a skeleton in two weeks. A body of a thin person, especially a child, decomposes slower than that of an obese person or an older person. In addition to rate of decomposition, the FAC scientists also carry out research on the bacteria colonising the body at different stages decomposition. It has been found that bacteria in the soil surrounding a decaying body can provide valuable information relating to the time of death.

Some bodies are placed in the open air and others in the woods. The effect of shallow graves, of leaving the body in a car boot or immersed in water, all play a part in establishing the mechanisms of decomposition. At the Anthropology Research Facility these different variables can be studied scientifically (Figure 7.1). The human subjects are those that have been donated to medical research. Although valuable information is being produced by the Facility by examining bodies in different conditions, we need to bear in mind that those conditions may be quite specific to Tennessee. At other locations there may be different factors that influence the course of decomposition.

7.21 EVERY CONTACT LEAVES A TRACE

Forensic science, the application of science to legal cases, is now firmly established as being one of the main tools for investigation of serious crimes. Although it is generally used in

Figure 7.1 In the Forensic Anthropology Center at the University of Tennessee they use bodies donated for scientific research to study how they decay under the different conditions of an outdoor environment.
Image courtesy of the Forensic Anthropology Center, University of Tennessee.

criminal investigations, it is not restricted to these, for it is also called upon in civil cases.

Despite the major role played by forensic scientists in providing the material evidence for criminal investigations, they are only a part of the investigating team. The skills and experience of others, such as the police, are just as important as they were before forensic science came on the scene. For example, it is the local police officer who knows the area and knows who the villains are. He has the experience to spot the *modus operandi* (MO) of a criminal and direct the investigations at that particular suspect. Thorough detective work frequently brings the culprits to justice with the minimum of delay, and before he has time to offend again.

Forensic science began in earnest in the 19th Century as a result of the rapid developments in chemistry. Results of chemical tests for detecting poisons were beginning to be considered as evidence in the courts. The trial of Marie Lafarge was a milestone case. In this trial, which took place in 1840s France, she stood accused of poisoning her husband Charles. The court case pulled together experts in toxicology to examine whether or not Marie had poisoned her husband with arsenic.

In this trial the results of an earlier arsenic test on the body of Marie's husband had been presented to the court. However, there was doubt about the reliability of the results and confusion over which method should have been used. To resolve the matter, the body of Charles Lafarge was exhumed some eight months after his death.

> *The coffin was little more than three feet below the surface, and when opened, the body presented a hideous spectacle, and was so much decomposed, that instead of the usual instruments, it was necessary, to use a spoon to take samples. The body was paste rather than flesh and was put into earthen pots to be brought for tests.*
>
> (***The Spectator***, **1840**)

The remains were then tested with an improved test developed by James Marsh (1794–1846) a chemist at the Royal Arsenal, London (Figure 7.2). Marsh had developed his test from the arsenic mirror test discovered by Johan Metzger in 1787. Metzger's test could provide only a qualitative analysis for arsenic, whereas

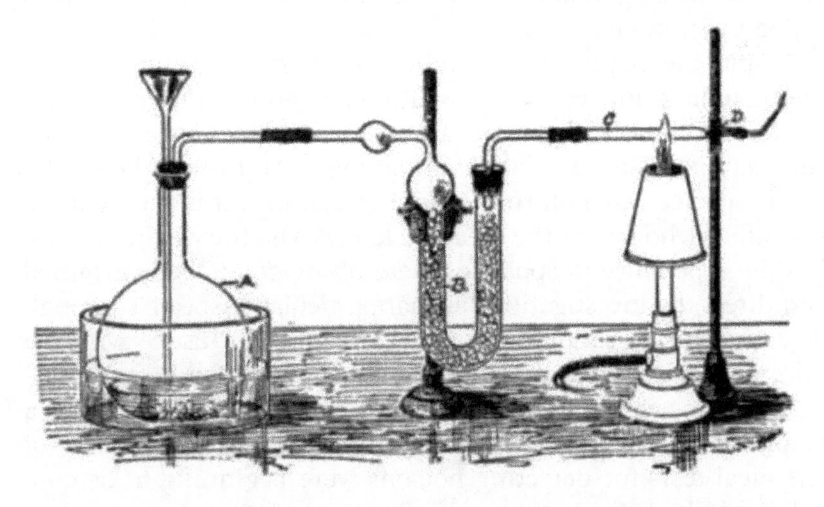

Figure 7.2 In the 1830s James Marsh developed a method for detecting arsenic in minute amounts. With suitable sample preparation his method could be applied to body parts and fluids from fresh corpses or decomposing remains. The results of arsenic tests using the Marsh apparatus played a vital role in the conviction of Marie Lafarge in 1840s France.

Marsh's test, reported in 1836, was quantitative and highly sensitive. There still remained questions about the analysis results, but what was demonstrated was the need for high-purity chemicals for the test method. In the arsenic tests, such as Marsh's test, the quality of the zinc used to generate the hydrogen was especially important, for zinc often contains a small amount of arsenic.

As we note from the piece in *The Spectator*, the taking of samples for forensic analysis can be, to say the least, unpleasant, especially if it involves a dead body. The following case is another example from the early days of forensic science and again involves the analysis of decomposing remains. It was the Lamson case and the trial was at the Central Criminal Court on 8th March 1882. Evidence was presented relating to tests carried out on the victim's remains. Thomas Stevenson was an analytical chemist and medically qualified.

He made extracts from Percy's internal organs, although by the time the jars were in his possession, the small intestine and colon were already somewhat green with decomposition. There was no specific test for aconite and so Stevenson relied upon tasting each extract and comparing the taste with that of laboratory standards of pure chemicals. Stevenson had about 60 preparations for this purpose. He tasted the extracts from Percy's stomach, urine, vomit, liver, spleen and kidneys. His tests indicated aconitine in the stomach contents, urine and vomit. Lamson was found guilty and sentenced to death.

Forensic science developed in France ahead of other countries. A central figure in this development was Edmond Locard (1877–1966) who set up a laboratory in 1910 at Lyons. Locard is perhaps best known for the phrase: "Every contact leaves a trace". He was inspired by Arthur Conan Doyle's *Sherlock Holmes* stories such as *Study in Scarlet*, which appeared in 1886.

Conan Doyle had qualified in medicine at Edinburgh University in the 1880s after submitting his MD thesis on syphilis. It is interesting to note that the poisoner Robert William Buchanan also qualified in medicine at Edinburgh about the same time. And only a few years before Lamson had attended Robert Christison's lectures on poisons at the same university.

Forensic science and the presentation of scientific results as evidence in court was being firmly established. It was not long before other tests were developed, such as fingerprints as contact evidence that represented a major breakthrough. In the following paragraphs we will see how a modern forensic investigation works its way through a series of well-defined activities: assessing the crime scene; recovering the evidence; examination of the evidence; analysis of physical evidence; and interpretation of the results. The fact that in this book we are studying poisonings makes no difference. The forensic techniques that apply are those used for other types of crime.

7.22 INCIDENT SCENE

In cases of poisoning, the scene is typically one in which there is a body present and the cause of death is murder, suicide or accident. But there isn't always a dead body, for in some cases of suspected poison the victim is still alive, despite being at death's door. And there are instances where murder by poisoning is suspected but the corpse has gone.

Forensic investigation of poisoning must focus upon the body, for this is where crucial information is to be found. Each case is unique and so there is no exact routine to work through. However, there are generally accepted ways of examining the incident scene and these are outlined below. The aim in examining a crime scene is to answer a set of questions, among which the more important are: When was the crime committed? Who is the victim and who is the perpetrator? What was the sequence of events? Why did the crime take place? Asking when the crime was committed is often the first question for the answer can often relate to other events that have a bearing upon the crime.

Despite appearances, the victim of a suspected poisoning may not be dead. Thus, the first step when a fresh body is found is often a medical examination and, if alive, provision of emergency medical treatment. Only then, can the forensic science begin. Of course, if the victim is taken away to hospital a major source of physical evidence has been removed and much associated evidence damaged in the process.

For investigating the incident, important first actions are for the Scientific Support Manager (SSM) to organise preservation of the scene in the state in which it was found. This involves restricting access to that of authorised personnel such as Scenes of

Crime Officers (SOCOs) and to organise a Common Approach Path (CAP) through the cordoned off area so that those attending the scene follow the same route, and do not go trampling all over the place. By such means, damage to evidence that may be scattered around the scene is kept to a minimum.

In managing an outdoor incident scene it is usual to arrange for some form of protection against the weather. The scene must also be protected to avoid the perpetrator returning to the scene, from the public who want to come in and nosey around, and from access of animals – especially if a dead body is at the scene. The SOCOs carry out an overview of the scene, make records of the scene, and organise photographs and videos. They may choose to call in the appropriate experts, for example, a forensic entomologist is needed if a discovered body is crawling with maggots.

A scene at which a murder victim is found, often does not necessarily mean that the murder took place there. The actual murder scene may be somewhere else as bodies are often moved – sometimes with great difficulty. Of course if the victim was killed as a result of chronic poisoning, there is not going to be a definable murder scene. An incident scene is a changing environment. This is particularly so when a dead body is the main focus. When a fresh body is found, information must be retrieved as a matter of urgency.

7.23 PHYSICAL EVIDENCE

Physical evidence at the crime scene may present itself in many forms. Important in the early stages of the investigation is the recording and preserving of marks and impressions left at the scene. For example, evidence arising out of contact between say culprit and victim is crucial information.

Locard's statement that every contact leaves a trace is a foremost consideration in all forensic work. For example, in a murder where the killer and victim have been in contact, the killer leaves traces on the victim and the victim leaves traces on the killer. Contact evidence is commonly present in the form of fingerprints, bite marks on a body, body fluids (blood, saliva, semen), footwear marks, tool marks and weapon marks. Recovering physical evidence, whether that be trace or bulk, is not always a straightforward task, for once that evidence is moved, such as in the sample collection process, some other evidence is lost or damaged.

Furthermore, handling the evidence may cause its contamination with adjacent materials or with traces from the person carrying out the sample collection. The risk of contamination is reduced by the forensic scientists, and anyone else involved, wearing protective overalls and caps. This guards against transfer of clothing fibres and hair. Facemasks may be used to protect against DNA from breath and sneezes being transferred.

The evidence is then packaged, labelled and stored pending examination by the forensic science laboratory. In England and Wales the forensic science laboratories were, up until 2012, owned by the government but are now run as private companies.

Physical evidence is labelled to identify each person who is responsible for the item from the time it is taken from the crime scene, to when it is finally destroyed at the end of the investigations. This sequence of recording is known as a 'chain of custody' and any person in that chain can be asked about the evidence while it was in their keeping.

7.24 BODY SAMPLES

Death by poisoning may show no outward signs that suggest foul play, and so the investigation of suspected poisonings calls for the internal organs of a body to be examined in a post mortem. This is where the expertise of the forensic pathologist becomes essential. The body is taken to the mortuary and given a thorough external inspection. There may be obvious signs of death and there may be the subtle clues that point to a particular cause of death, and which may be noticeable only to the expert eye.

Using a scalpel, the pathologist makes a Y shaped cut from the shoulders to the pubic region. This enables access to the internal organs and, after cutting the ribs, the major organs of the thoracic and abdominal areas are removed. Samples of body fluids are also collected during this process. Removal of the brain from the head needs very careful handling to prevent damage. A saw is used to produce a circular score around the top of the skull and then a cranium chisel is required to gently open the cut. The top of the skull is lifted off and the brain removed.

7.25 COURT APPEARANCE

After all the detective work and analysis of results, a man stands before the court. When the powers that be believe they know truth, the man is brought before the court. Although the court may think they know the true facts there is always an element of doubt. As such there is always a trace of subjectivity. Thus, a wise measure is for him to have a haircut, wear his best suit, shirt and tie and, if pleading guilty, to show remorse.

CHAPTER 8

Chemistry Clues and Crime

In days gone by, once the poison was absorbed by the victim's body it was extremely difficult, and in most cases impossible, to detect. There are many poisons that give, to all outward appearances, the impression that death occurred through natural causes. Things have changed. Today there may well be more poisons available to the individual than ever before, but there are also advances in medical examination and forensic analysis that increase the likelihood of the poisoner being caught.

Analytical chemistry plays a major role in forensic science, and we must therefore know something of the main techniques and how different poisons call for different analysis procedures. Analysis may be performed on a wide range of materials taken from a body itself or from items from the incident scene. The aim is to identify a suspicious substance (qualitative analysis) and maybe to measure the amount of that substance (quantitative analysis).

Quantitative analysis is especially important in the investigation of poisonings. Tissue from a dead body may well indicate the presence of a poison, but the amount present may be sublethal, and it may be naturally present in concentrations known as the 'background level'. For example, arsenic may be detected in the hair at a level that is natural, because there is a certain amount of arsenic in the environment and we all absorb it to a

Poisons and Poisonings: Death by Stealth
By Tony Hargreaves

Published by the Royal Society of Chemistry, www.rsc.org

minor extent. It would be important from an evidence point of view if the quantity of arsenic in the forensic sample was significantly higher than the background level.

Of course, it would be foolhardy and possibly embarrassing to report that the poison arsenic had been found in a dead body, and possibly have that statement taken as evidence of poisoning. At some point in the investigation some bright-spark might decide to intervene and say that this was nothing abnormal as we all have arsenic in us. The question then would be: How much arsenic is normally present? A detailed examination of the analysis data would then follow.

Where quantitative analysis is required to determine the amount of a particular poison, it is usual practice to perform a series of repeat determinations along with appropriate controls, standards and calibrations. With several results, a statistical treatment can be used to provide a high degree of confidence in the results and the conclusions that are based upon them. Thus, precision and accuracy become important here. Precision is the smallest difference that can be detected in a measured quantity. Accuracy is how close the results fit the true figures.

Rarely does a chemical present itself in pure form that can simply be sampled and dispatched to the laboratory. Often the chemical is concealed within the dead body, dissolved or dispersed in tissue or body fluids. In other instances it may be outside the body and mixed with other substances, or adhering to materials where it may be visible to the naked eye or by means of a hand lens. For example, in a case of suspected arsenic poisoning a range of samples may need to be tested: stomach contents; tissue samples; hair samples; body fluids; white residue from the bottom of a wine glass; a smear of white powder on clothing; specks of powder on a work top sampled by means of adhesive tape (lifting tape); and even air samples if the gaseous compounds of arsenic are suspected. To transfer the material to be investigated from the crime scene to the laboratory is going to require appropriate sampling methods, followed by preparation of the sample to suit a particular line of analysis.

The results of any analysis are only ever as good as the sample itself, and so the process of sampling and preparation must be carried out with the utmost of care to ensure a high-integrity sample. If the integrity of the sample is in doubt then the results are worthless. Also, a chain of custody must be preserved with all those responsible for handling the sample signing the accompanying label. There must be no break in the chain. Thus, the risk of tampering with the sample, or dishonestly switching it for another sample, is reduced. If there is any chance of the sample being handled by an unauthorised person, the sample can no longer be relied upon as physical evidence.

8.1 TRACE OR BULK ANALYSIS

In much forensic work the amount of sample available for analysis is likely to be minute. It might be a single human hair or some specks of white powder collected with lifting tape, or even *Sellotape* itself. The sample is present in trace amounts and there may be sufficient for only a minimum number of tests.

On the other hand, there may a relatively large amount of sample available for analysis, as might be the case where a 100 g jar of unknown white powder had been discovered in the kitchen cupboard. In this case, the analysis would be a bulk analysis. The laboratory performing trace analysis is often different from the one carrying out bulk analysis, to avoid the risk of cross-contamination from bulk sample to trace sample. Trace samples are highly susceptible to contamination, and can easily be rendered useless. For example, a tiny amount of contamination introduced to a sample that is itself minute, represents a high proportion of contamination.

8.2 SIMPLE, RAPID TESTS

Not all tests call for a fully equipped laboratory with the latest in analytical instruments. There are many simple tests that can be carried out quickly, and sometimes at the incident scene, which give an indication of what direction the investigation should move in. For example, consider what steps would be taken if specks of white powder aroused suspicion. Some common possibilities come to mind. Could the powder be, say, cyanide? Is it a single pure chemical, or a mixture of white powders, such as cyanide mixed with salt? Carrying out some simple tests will give a good idea.

Appearance and odour would be noted as a first step, and this may provide clues. For instance if our white powder smells of almonds the immediate thought would be that it is (or contains) cyanide. Further tests could be organised, starting with the test for cyanide. This could be done quickly, at low cost and without the need for sophisticated instruments. Test kits are available to aid identification. They would be used only when there was ample material, as they involve mixing the sample with chemicals to produce a colour reaction. In the process the sample is destroyed.

8.3 NON-DESTRUCTIVE TESTS

As a first action, it is always the case to carry out tests that do not destroy the sample. This is essential where only minute amounts of sample are available. As much of the evidence as is reasonable must be retained. Some tests are non-destructive and leave the sample unaffected and available for further examination.

Of course the first examination will involve appearance and smell, and can be noted before the sample is even touched. For example, in a case of a suspicious death of a young child, the stomach contents were being examined visually. In the initial visual examination some red shell-like bits were found. The idea that these were the skins from poisonous red berries came to mind, and would have been consistent with poisoning. It is a fact that children are attracted to bright red berries. They pick them, taste them and sometimes they swallow them. The shell-like red bits were picked out with forceps and examined more closely and found to be the gelatine shells of capsules used for many medicines.

It turned out that the child had not swallowed poisonous berries, but had taken someone's pills and received a fatal dose. The pills were the gelatine capsule type with bright-red colouring. When these break open to release the active ingredients they leave the shell-like gelatine, which dissolves only slowly. Had the child swallowed the pills and died accidentally, or had someone given them to the child. Accident or murder? Crucial evidence had come to light simply by careful visual inspection. Quick results were obtained, without the expense of chemical analysis and without destroying any of the sample.

Clearly, non-destructive tests must be exhausted before other tests are employed. Many tests are destructive, the sample becomes consumed in the process of analysis, and if something goes wrong in the analysis evidence is lost. In the example of the white powder referred to above, it would be sensible to start with a visual examination. Using a hand lens we see if the particles clump together as if damp. Or are those particles free and easily separated? Sodium cyanide is crystalline and the crystals tend to cling, due to moisture absorbed from the air.

Placing the powder under the microscope and examining the shape and size of the crystals would extend the visual examination. The sample can be removed from the slide and used in further tests. Examination by this method will show if the powder is a mixture by revealing different crystal shapes. There are sophisticated variations of microscopy that can provide much information and leave the sample unchanged. Furthermore, methods such as X-ray fluorescence are powerful, precise and leave the sample intact and available for further tests.

However, many tests use up the sample in the course of preparation and analysis. Such tests are used after all non-destructive testing options have been exhausted. For example, in preparing a sample of body tissue to produce a solution for atomic absorption spectroscopy, for heavy metals analysis, the tissue is digested in boiling acid until all the organic matter has been broken down. There may be enough solution for several runs through the instrument, but there is no going back to that little bit of body tissue.

8.4 ANALYSIS

Identifying a poison and measuring its concentration in human organs calls for a range of analytical methods, the more important of which are outlined below. Analysing poison from a dead body is made more difficult by the presence of the body's decomposition products, which can cause interferences, and the fact that the poison may have itself broken down and changed into other substances. These problems occur especially with exhumed remains where there is the additional problem of chemicals being absorbed from the surrounding soil.

Analysis of poisons from a living person is not going to be compromised by the presence of decomposition products, and the poison may still be in a reasonably fresh condition. However, with a living person, the option of examining the internal organs is ruled out.

8.5 MARSH'S ARSENIC TEST

This test is no longer used in forensic science, but it deserves a place here as it was a landmark in forensic science. It introduced quantitative analysis of arsenic in body fluids and organs. From the 16th Century onwards, when arsenic became the common method of poisoning, there was a need for a reliable test that could establish its presence in the human body.

Prior to the Marsh test there was a crude test to see if a white powder was arsenic. If the powder was put onto a charcoal or coal fire and released a smell of garlic, it was taken to be arsenic because this is what arsenic does. The test was crude and the interpretation of the result was crude because it was a generalisation in which no account was taken of the fact that there may be other chemicals that can produce a smell of garlic.

A better test for arsenic was to prepare a solution of the powder and bubble hydrogen sulphide through it. A yellow precipitate was formed if arsenic was present, and further tests could be carried out on the precipitate to confirm that it derived from arsenic. In the case of Marie Lafarge, this precipitation test was used on samples from her dead husband and what he had recently ingested. Thus, arsenic tests were carried out on egg nog, soup, sugar water, vomit and stomach contents.

James Marsh based his test upon the arsenic mirror reaction first reported by Berzelius and later refined by Metzgar. The principle was to react an aqueous preparation of the sample with hydrogen produced from zinc and hydrochloric acid. The hydrogen, along with arsine gas from the arsenic, was passed into a heated glass tube, wherein the arsine formed a mirror. By carrying out the test on suitable known samples and blank runs, Marsh was able to obtain quantitative results – although by today's standards these would be regarded as only semi-quantitative.

Marsh also described the methods suitable for treating organic material from a corpse to make it suitable for the arsenic analysis. For example, stomach contents needed to be digested in strong acid to destroy any organic matter before the arsenic test could be applied. The Marsh test of 1836 was successful in demonstrating the value of chemical analysis in providing sound evidence for the courts. However, it was soon followed by an arsenic test devised by Edgar Reinsch in 1842. The Reinsch test became popular, as it was portable and could easily be demonstrated in court. Furthermore, it was sensitive enough to detect extremely small amounts of arsenic down to 0.0000001 g arsenic in the sample. Both the Marsh and Reinsch tests were instrumental in providing reliable evidence on which many accused of poisoning were judged innocent or guilty.

The above tests are no longer used in the forensic examination of poisoning cases. Today we have, at the leading edge of arsenic analysis, the non-destructive technique of neutron activation analysis. This can detect arsenic at one particular spot within a single human hair and without even changing the sample.

8.6 THIN LAYER CHROMATOGRAPHY (TLC)

A separation technique used in forensic science laboratories to isolate substances for subsequent analysis. Example: stomach contents from the body of a person suspected of being poisoned or drugged are subject to TLC and then further analysis. The sample must be prepared to provide a solution, either aqueous or in an organic solvent.

The prepared liquid is spotted onto a thin layer of coherent powder supported on a glass or aluminium plate. Silica gel powder is commonly used and the plates are usually square with sides of 15 cm. A small volume of solvent mixture is placed in a tank and the plate stood in this such that the solvent slowly creeps up the plate by capillary action, and reaches the spots.

As the solvent front travels up the plate it passes through the sample spot and carries with it the chemicals that are present in the spot. The silica gel attracts different chemicals to different extents and thereby separates each. When an adequate separation has been achieved, the plate is removed from the solvent tank and allowed to dry. By proper choice of conditions, a good separation is produced as the solvent front nears the top of the plate.

The dry plate is then sprayed with reagents that produce a specific colour for a particular chemical. By running the test in parallel with appropriate standards, the spots representing different components in the sample can be identified. Sometimes the spots are visible when viewed under ultraviolet light.

8.7 GAS CHROMATOGRAPHY (GC)

Poisonous organic compounds and drugs are often analysed by another type of chromatography known as 'gas chromatography'. The material fed into the instrument must be volatile as the analysis works by evaporating the sample and then separating the components of the vapour. Not all samples are volatile, and so they must first be converted to a volatile derivative.

The prepared liquid sample is injected through a septum by means of a syringe into the heated chamber where it evaporates. An inert carrier gas such as helium mixes with the vapour and transfers it to a coiled capillary tube known as the 'column'. This tube is coated on the inside with a wax such as silicone and is maintained at an elevated temperature in the precision oven. In the vapour mixture the different substances are attracted to the wax to different extents, and so become separated as the sample vapour passes along the column, which is typically 2 m in length.

As the separated substances emerge from the column, they enter a detector, usually a flame ionisation detector. This produces an electrical signal to be recorded in a computer or chart recorder. Each separated compound is shown as a separate peak on a chromatogram, and the area under each peak is proportional to the amount of substance. Identification and quantification of each emergent compound is achieved by comparing the result with calibration samples containing precise amounts of known chemicals.

Another means of identification is to have the GC coupled to a mass spectrometer (MS), described below, in an arrangement referred to as 'gas chromatography-mass spectrometry' (GC-MS). In GC there are other ways in which the sample vapour can be introduced to the column, such as headspace GC and pyrolysis GC.

8.8 MASS SPECTROMETRY (MS)

Identification of many different substances can be carried out by MS. Some instruments are designed to take solids, whereas other instruments can accept the sample in liquid form or as a vapour. The MS cannot separate mixtures of chemicals and so it is often used in conjunction with GC. The GC separates the components of the mixture and then feeds these into the MS for analysis.

The separated chemicals from the sample vapour are converted to positive ions. These ions form an ion beam that is accelerated into a magnetic field that deflects the beam to an extent that is proportional to the mass of the particular ion.

Changing the strength of the magnetic field enables ions of a particular mass to enter the detector. As with GC, the detector feeds the information to a chart recorder or computer. Two main types of information are provided: molecular mass; and fragmentation pattern. From this, along with published tables of data, it is possible to identify the molecules passing through the MS as they leave the GC section.

8.9 HIGH-PERFORMANCE LIQUID CHROMATOGRAPHY (HPLC)

In principle this method is like the GC procedure described above. In practice, it works on samples that are prepared as a solution. When placed in the instrument the liquid sample is carried through a separation column by a flow of carrier solvent. The mixture becomes separated into its component parts, which then emerge in a series and are carried to a detector, such as a UV detector, which produces a signal for the recording device. Identification is by means of calibration solutions containing known amounts of the relevant chemicals.

8.10 ATOMIC ABSORPTION SPECTROSCOPY (AAS)

Many suspected poisonings involve metal compounds. Some notable examples are the compounds of heavy metals such as mercury, lead and arsenic. Being able to detect and measure these in minute concentrations, down to a few ppb, is a vital part of modern investigations of poisoning. In AAS this is achieved relatively rapidly, although the sample preparation is often a slow process.

The sample must be prepared as an aqueous solution in which the metals are dissolved so that they are present as ions. Where the original sample is flesh, stomach contents or food, the procedure involves boiling the sample in strong acid to destroy all the organic matter, but leaving the metal ions in solution.

In the instrument, the sample solution passes into an atomiser where it is converted into a fine mist. This mixes with a fuel (often acetylene) and air mixture and flows into a burner where a high-temperature flame is produced. Metal ions present in the sample solution are converted into gaseous metal atoms. A special type of lamp known as a 'metal cathode lamp' produces light that passes through the flame. If analysis for lead is being performed, a lamp using a lead cathode is used. This produces light of a frequency specific to lead, so that lead atoms in the flame absorb it.

The remainder of the light passes to the detector, which sends a signal to the chart recorder or computer. The amount of light absorbed is proportional to the amount of lead in the solution. By using standard lead solutions for calibration, the instrument then gives a figure from which the lead content in the original sample can be calculated. One particular lamp enables analysis for that one metal. Where several metals are of interest, different lamps may be needed.

Some analyses require flames of even greater temperature than that provided by the acetylene and air. As such, the choice of gas mixture is decided by the particular analysis. For some applications, an electric graphite furnace replaces the gas flame.

8.11 INFRARED SPECTROSCOPY (IR)

This is used for identifying mainly organic compounds so long as they are pure and dry. The sample may be prepared in a variety of ways: as powdered solid compressed into a disc with potassium bromide; as powder, mulled with pure mineral oil; or as solid dissolved in an organic solvent or a liquid. Inside the instrument a beam of infrared light is projected through the prepared sample, and the emergent beam passes into a detector. The amount of light absorbed by the sample at various wavelengths is detected, and a signal passed to a chart recorder or computer.

The light source provides light over a range of wavelengths, which enables the sampled to be scanned to produce several absorptions at different frequencies across the whole of the infrared spectrum. A very useful feature of IR is that it provides a unique spectrum, known as a 'fingerprint', for each and every compound. Reference spectra are published for a huge number of organic compounds to facilitate identification of an unknown substance.

8.12 X-RAY ANALYSIS

X-Ray fluorescence (XRF) is a non-destructive method of analysis, and so the sample is unaffected by the test and remains available for further investigation. The method is capable of identifying and measuring a large range of elements in a solid sample, and involves directing a beam of X-rays of a particular wavelength at the sample. This causes the sample to fluoresce and emit X-rays of a different wavelength to the original, which are characteristic of the element causing their emission. A detector measures the X-rays and sends a signal to a recorder.

In X-ray diffraction (XRD) the sample present as a powder is analysed. The powder is composed of tiny crystals. An X-ray beam is aimed at the sample and when it strikes atoms in the different layers of the crystal lattice, some of the X-rays are reflected when a critical angle is reached.

The reflections are recorded on photographic film and measurements made to determine the spacing between different layers of atoms at different orientations. From this, the chemical structure of the substance can be worked out.

8.13 NEUTRON ACTIVATION ANALYSIS (NAA)

This is not a routine technique, but was used for the analysis of a sample of Napoleon's hair. The sample is subjected to a neutron flux. When a neutron collides with a target nucleus, the neutron is absorbed to form a compound nucleus. This undergoes changes and emits a gamma ray having an energy that is a characteristic of the target nucleus.

Using NAA produces results at very low detection limits, uses minute amounts of material, *e.g.*, a single hair, and is non-destructive. At the end of the analysis the sample can be removed from the instrument and used for other analyses.

8.14 DNA AND ELECTROPHORESIS

Although not directly involved for investigating poisons, the revolutionary method of DNA analysis deserves to be mentioned, for it is now a firmly established part of forensic analysis. It can help to identify the body of a victim through a blood relationship. Where there are several bodies in a mutilated state, it can ensure that the correct bits end up in the correct coffin. The perpetrator of a crime may also be found through DNA analysis and reference to the DNA profiles of suspects that are held in the National DNA Database (NDNAD).

DNA (deoxyribonucleic acid) is a molecule that contains the individual's genetic code, which carries information provided by both the mother and the father. Most of the cells in the human body carry DNA in their nuclei and mitochondria. Nuclear DNA is made up from genes from both mother and father, whereas mitochondrial DNA has genes from the mother only.

Each individual has a unique DNA code. Analysis of DNA involves extracting it from the sample, which may be only a minute amount of material. The amount of DNA extracted would be too small (down to a billionth of a gram) for analysis, but it can be amplified using PCR (polymerase chain reaction) to artificially increase the amount.

Once enough DNA has been produced, analysis is carried out using electrophoresis. In this procedure fragments of the DNA molecule are separated by means of an electrical field to produce a sequence of bands that correspond to parts of the molecule. The profile formed is unique to an individual's DNA and can be compared with profiles from the NDNAD.

8.15 SPOT TESTS

A suspicious substance may be tested using colour indicator reactions. These are convenient, cheap and easy to perform, which makes them useful for doing tests at an incident scene and producing a quick result. However, they have disadvantages in that they are only semi-quantitative and results from them must be backed up with further analysis at the forensic laboratory. Despite this they can usefully narrow down the range of possibilities. A major use of spot tests is in identifying substances thought to be drugs.

Examples of spot tests are: Marquis', Mecke's and Froehde's tests for opiates; Marquis' and Simon's tests for amphetamines and methamphetamine; Scott's test for cocaine; and Duquenois' test for cannabinoids. All these rely upon colour changes, for instance Marquis' test gives a dark-purple colour for diamorphine, whereas cocaine produces a pink/orange to indicate this drug.

Strychnine is one of the cruellest of poisons, and was one of the first to have a colour test developed for it. The test used sulphuric acid, potassium dichromate and manganese dioxide. In the presence of strychnine this reactive mixture gives a deep-purple colour that gradually fades to red. This colour reaction was introduced in the 1860s, before analytical instruments became available. At that time, a positive result for the test was taken as proof. Today, we might ask more questions such as: could other alkaloids give a similar result? Of course, we now have the means to draw on other tests that are reliable, accurate, precise, and can measure the smallest amounts of alkaloids. The methods are HPLC, TLC, IR, UV and GC-MS analysis.

Prior to the colour tests for alkaloids, there were precipitation tests, as most alkaloids form an insoluble substance when reacted with an aqueous solution of mercuric chloride and potassium iodide. The nature of the precipitate could reveal further information. In some cases, the precipitate would be fed to an animal to assess its toxic properties. Using frogs to test for strychnine was common. In the case of Palmer, a live frog was placed in strychnine solution. Standard solutions of strychnine were used and the physiological response and reaction time noted.

8.16 ROADSIDE CHECKS

These are now being used by the police for apprehending drug drivers. The breathalyser has been highly successful in tackling the problem of drink-driving, and now the drugalyser is the latest weapon against people who drive while intoxicated.

The breathalyser has also shown its worth in testing airline pilots who have been on the booze. A Latvian pilot was sentenced to six months in prison after failing a breathalyser test just before he was due to take off with a hundred passengers from Oslo airport bound for Crete. Further tests to determine his blood alcohol level showed he was seven times over the limit.

Whereas the breathalyser uses a fixed volume of exhaled air, the drugalyser works on a sample of saliva from under the suspect's tongue. The saliva passes into a nitrocellulose strip containing antibodies corresponding to immobilised drugs, and causes them to change colour. Drugs such as cannabis, cocaine, ecstasy, heroin and methamphetamine can now be detected and the amounts measured.

8.17 POISONED AIR

Detecting poisonous substances in the air is called for in many situations. For example, in coal mines, radon in buildings, carbon monoxide in the home or caravan and poisonous gases, fumes and dusts in the working atmosphere. Colour indicator tubes are commonly used for gas tests, but filter methods must be applied for fumes and dusts (Table 8.1.).

8.18 RECONSTRUCTION AND SIMULATION

In attempting to identify a body that has suffered the effects of decomposition, forensic sculpture is valuable. This normally involves the skull and uses modelling clay to build it up where the flesh would have been. The sculptured final form gives a good impression of what the person looked like, and may be

Table 8.1 Indicator colour tubes used to test for the presence of poisonous gases and vapours in the air.

Gas tested for	Type of tube	Colour change
Arsine, arsenic hydride	Gold	White to grey violet
Carbon dioxide	Crystal violet and hydrazine	White to violet
Carbon monoxide	Iodine pentoxide	White to brown
Chlorine	Tolidine	White to orange
Ethanol[a]	Chromium(IV)	yellow to green
Hydrogen cyanide	Mercury(II) chloride and methyl red	Yellow to red
Hydrogen sulphide	Mercury	White to brown
Mercury vapour	Copper(I) iodide	Pale yellow to orange
Nitrogen dioxide	Diphenylbenzidine	Yellow to grey
Ozone	Indigo	Blue to white
Sulphur dioxide	Iodine	Blue to white

[a]This tube is similar to the tube used in the early roadside breathalyser.

photographed for the public to see in the press or on TV programmes such as the BBC's *Crime Watch*.

The biggest reconstruction job ever undertaken involved not a single victim, but the remains of the aircraft in the Pan Am 103 disaster that turned out to be Britain's biggest mass murder with hundreds of victims. As the aircraft broke up in the sky, there were bodies and body parts, along with aircraft parts scattered over a vast area. In fact, bits were found over an area of 850 square miles. Eventually there was sufficient material collected to enable 70% reconstruction of the aircraft remains in a large hangar. Efforts to put the body parts of individuals together were assisted by various methods.

8.19 WHAT DO THE RESULTS MEAN?

We must be aware of the quality of the results that come from any analysis. The quality of any result depends upon the quality of the sample: poor sample, poor result. The process by which forensic results are obtained must be rigorous, with attention to aspects of good laboratory practice such as standards, calibration and instrument checks, and blank runs.

Where analysis for a poison has been performed, the results often give the amount of the poison in the sample, for example, the weight of barbiturate in the deceased's liver. The actual figure is vital to the conclusion, and this puts the emphasis on the numbers themselves. The amount of barbiturate in the liver might have been determined in different parts of the organ and a scatter of results obtained. There may also have been repeat determinations in the same part of the liver to act as a check on the reproducibility of the method.

Several figures may be produced, and if those figures are near to the values that would indicate lethal concentrations, they must be examined in some depth. As such, statistical treatment of the results may be called for. In its simplest form this would provide an arithmetic mean (the average value) and the standard deviation (the spread of values) for those figures that cluster together.

Precision and accuracy are obviously important, but we need to be clear of the meaning of each in order to take a critical look at the figures. This is best appreciated by considering a dartboard. If one player succeeds in placing his darts around the bullseye, but not particularly close to each other we would regard this as 'accurate but not very precise'. If another player placed his darts all very close together but at a point that was some distance from the bullseye, then we would say that his dart throwing was 'precise but not accurate'. Of course, the ideal is to have both accuracy and precision.

The examination of data can take us into some complex treatments that require the skills of a statistics expert. Examination of data is aimed at working out the probability or likelihood of an event. This can be a demanding exercise involving probability, odds and likelihood ratios, which are calculated by means of the statistical equation known as Bayes' Theorem.

8.20 PROBABILITY

This is a measure of how certain we can be about an assertion being true. Here is a simple example. A witness might say that the person observed at the crime scene was male. We might challenge this, but there is no doubt that the person was either male or female. The probability of an assertion being true is assessed on a scale based on zero to one, where zero represents the assertion is false and one represents the assertion being true. Sometimes, percentages are used for quoting probability. We would say that the probability of the person at the crime scene being male is 50%.

The forensic scientist will have formed an opinion based upon the most probable explanation of the results, and this will be presented in a report. The scientist can express an opinion using a likelihood ratio. This helps the court to establish the odds of the accused person being guilty. However, such a task can be far from straightforward. There are mathematical formulae for working out likelihood ratios, but these are involved, and beyond the scope of this text. Having an understanding of probability can go some way in understanding faulty reasoning and fallacies.

8.21 SEQUENCE OF EVENTS

Piecing together the sequence of events of a crime, invariably leaves gaps. The net effect is a compilation of facts and gaps – holes in the fabric of knowledge. These holes become filled with belief. It is logical that . . . . This must have taken place . . . . Surely he would have . . . . Facts are objective and provide knowledge. Belief is subjective and creates a story. Occasionally, confidence in the results is destroyed if it comes to light that the material under examination at the laboratory was contaminated or tampered with.

The perpetrator of a crime may unwittingly fill in some of the spaces. Here we consider the Soham murders in which the killer was Ian Huntley. If he had suddenly disappeared after committing the crime then suspicion would, quite naturally, have fallen upon him. But he stayed within the community and behaved normally, assisting the police just as any other person would. In the early part of the investigation there were many gaps in the pattern of events. No one knew the full story, that is, apart from Huntley.

In an effort to show interest and concern, he asked one of the investigators if the girls' clothes had been found yet. This was picked up on as being highly suspicious. Whoever mentioned the girls' clothes? Huntley had blown it. In his attempts to keep himself up-to-date and to talk about the events just like anyone else, he let something slip. The point here is that the killer knows more about the crime than anyone else, and is very likely to let something out in the course of conversation. The alternative would have been to remain silent about it but, that in itself, would have been cause for suspicion, especially for Huntley because he knew the girls well. A killer recognising that he has 'blown it' may come clean in the hope of reducing the severity of the sentence. Much expensive forensic work and police time was saved by him offering those few words.

8.22 POISONER PERSONALITY

At the crime scene there may be clues as to the behaviour and personality of the perpetrator, which may enable an analysis of the psychology of the poisoner to be worked out. Psychological profiling of those committing serious crimes, in particular, crimes of violence, was developed by the FBI in the 1960s. The research for this was based upon interviews with criminals, and the aim of the exercise was to assist in catching those involved in violent crimes. The question posed was: what sort of person are we looking for?

Much of the research focussed upon offenders who repeatedly raped or killed. It appeared they were driven by the high level of public fear that their crimes produced. Examination of the offenders' habits revealed similarities between one crime and a previous crime. For many offenders, it seemed that their serious crimes began in their 20s and were driven by the need for a sense of power over their victims, and a desire to manipulate them. A primary consideration in working out a profile is what drew the criminal to the victim, and what was their relationship with each other? A profile can reveal details such as personality traits, marital status, and whether employed or unemployed. Profiling is especially successful where the criminal has some form of mental illness.

Some criminals take away a trophy from the crime, some carry out a particular form of mutilation of the body, and some leave their 'trade mark' by, say, placing the corpse in a particular orientation.

The results of behaviour studies when made public, have been successful in people recognising the details as being those of someone they know, becoming suspicious of an individual, and contacting the police. For example, the profile may indicate: sex; age; occupation; race; or lifestyle.

In producing a profile for a serial criminal, there are two basic approaches. 'Inductive profiling' assumes the criminal has a background similar to that of other serial criminals who have behaved in a similar manner. This, of course, involves generalisations being made, and the technique has been criticised for this very reason. 'Deductive profiling' is the other approach. Here, generalisations are avoided and the profile is built upon specific details in the criminal's actions before, during, and after the crime. Research into the methods of working and the tools/weapons used, produce valuable information relating to behaviour, personality and background of the criminal.

Studies show that there are three main groups. The first is where the criminal is organised. He has a plan: he brings tools or weapons and removes them after the crime; he takes care to leave no evidence; and he hides or disposes of the body. In this group we find intelligent people with stable lifestyles, who are married and employed.

In the second group, we have the disorganised criminal. He leaves a mess: he does not bring tools or weapons; he lives alone or with a relative, often his mother or sister; he is unemployed; he has low intelligence or mental illness; and his attacks show extreme violence. The third group combines features of the previous groups. The criminal in this group is organised, but his pattern of attack is frenzied, out of control, and driven by deep-seated urges and fantasies.

In examining why a person behaves in one of the above ways, it is often found that they have mental damage from when they were young children, due to something a parent (or other adult) did, or as a result of some disturbing experience. Alternatively, some may have an inborn physiological abnormality of the brain that causes certain types of behaviour. Most of us, by the time we reach adulthood, are likely to be carrying some damage that was seeded in our formative years. There is a spectrum that goes from highly damaged to undamaged, and we all fall somewhere between the two extremes. Many of us may have fantasies that we would wish to indulge in, but we keep them under wraps, so to speak, and accept that there is a line that must not be crossed. That line is the law, and intelligent, well-balanced people know not to cross it.

Clearly, both nature and nurture can play a part. We should not read too much into these profiles. To believe that damaged people are destined to become criminals and undamaged ones will not become criminals would be wrong. But where does all this place the poisoner? And is there any psychological difference between the serial poisoner and the mass poisoner? The ideas may be sketchy, but even a vague profile can give an idea of motive.

To carry out a poisoning requires a knowledge of poisons, how to obtain them, and how to administer them. Clearly, a considerable amount of planning and organisation is called for, and this is especially so where the killer relies upon chronic poisoning.

Typically, the poisoner is either intent upon killing his victim, or simply wants to cause them non-fatal suffering as a means of punishment.

The motive to kill often includes at least one of these: money; love; power; or revenge. However, there are some interesting cases such as the so-called 'toxicomaniac' Graham Young, who used poisons to kill or punish. There seemed to be two separate motivations. His behaviour profile showed an obsession with poisons. He experimented with different poisons and different dose levels and meticulously recorded the results. To him, the victims were merely guinea pigs.

It is interesting to look at some cases to see what may have caused someone to become a poisoner. When considering these cases it should be born in mind that these poisoners were the ones that were caught, and didn't get away with it. It stands to reason that there are many in the past who were never caught, and there are some at work now who will succeed in getting away with murder. The details that follow relate to those who failed and paid the penalty. Despite these failures there must be many successes.

John Reginald Christie (10 Rillington Place) was a murderer who used poison to render his victims unconscious prior to killing them by strangulation. As such, Christie was not what we would usually regard as a poisoner. Before he started on a series of murders, which included at least six women, he had served prison sentences for theft and violent behaviour.

Christie grew up in Halifax in a family that was dominated by his father, who was a strict disciplinarian. His mother was over-protective. He enjoyed church, cinema, scouts and school. At the age of 11 he won a scholarship to high school, where he was found to be good at maths, and had an IQ of 128. When he was eight he saw the body of his grandfather in an open coffin, which had a lasting effect upon him.

He grew up with a dislike for women and seemed not to have had normal sexual relationships, which led to him being taunted about being inadequate. Although he married at the age of 19, his sexual problems continued, and he had a lifelong problem of impotence. He was an infantryman in World War I, suffered a mustard gas attack, and later he joined the wartime police service.

The type of employment taken up by poisoners may be significant. We note that there have been many poisoners in the caring professions, in particular in medicine. To name but a few: Dr Harold Frederick Shipman; Dr Thomas Neill Cream; Dr George Henry Lamson; Dr Hawley Harvey Crippen; Dr Michael Sevango; Dr William Palmer; Dr Robert Buchanan; Dr John Bodkin Adams; Dr Edward William Pritchard; Beverley Gail Allitt (nurse);

Donald Harvey (medical orderly); Colin Norris (nurse); Dorothea Nancy Waddingham (nurse); and Kristen Heather Gilbert (nurse).

8.23 CONCLUSION

As we close, we need to pull poisons, poisoners and poisoning together so that we leave with a balanced view. Poisoning has always been with us in one of its many forms. Most of today's poisonings are accidental. Old people become confused over their medication and inadvertently overdose. Children are poisoned by chewing or swallowing coloured berries, just as they have always done. Nowadays, there is the problem of children swallowing household chemicals, prescription pills and recreational drugs, and button batteries. Most of these poisonings cause pain and distress, but have no lasting ill-effect, so long as medical treatment is soon obtained.

There are also the accidental poisonings caused by our natural environment, and we are powerless to do anything about them. Poisonings occur due to industrial releases of dangerous chemicals. At first sight, we might label these as 'accidents' but on scrutiny we find many are due to criminal negligence, greed and corruption. People working in certain industries are prone to chronic occupational poisoning, as they are repeatedly exposed to poisonous chemicals, but this is now decreasing due to new safety regulations in many countries, but not all. Intentional mass poisoning due to chemical weapons rears its ugly head from time to time, despite international efforts to bring an end to their use.

Suicide is a curious one. Ending one's life because it is unbearable is commonplace in the so-called 'developed World' of 2016. It is odd that we have built a society in which all our primary needs are provided for, and many of us enjoy a relatively easy and comfortable lifestyle but, for some, death seems to be a better option. Poisoning as a means of suicide is as popular now as it was all those years ago when the Roman author Pliny the Elder (23–79 AD) noted the use of poison to relieve someone from the burden of living when life became unbearable.

It would seem that our high standard of living is, for many, failing to provide a high quality of life. The wealth that is so good at providing a life of pleasure, is not so good at offering a life of happiness. A large proportion of people seek an escape from the kind of living that modern society has imposed upon them. Perhaps it is something to do with the human psyche; our society is materially rich but spiritually destitute; those who feel the despair of it opt for the exit sign.

We live in an age of communication. Knowledge of poisonings, especially the more sensational ones, soon becomes global news. As a result, we have copycat poisonings where someone learns the details of a poisoning case and decides to try it out for themselves. This happened in the Tylenol poisonings, which set off a craze for product tampering and introducing poisons into foods and drugs on sale to the public. The intention was to kill or to hurt, but where the victims were unknown to the poisoner.

Money, power and love (or hate) come top of the list of motivating factors for deliberate poisoning, just as they did millennia ago. So what has changed? Today we have more poisons than ever because of the number of medicines we have access to. However, this is countered by the fact that medical examination and forensic science are highly developed, and establishing the details of poisoning generally succeeds in bringing the culprit to justice.

So what of the future? The classic trio, arsenic, cyanide and strychnine are unlikely to figure to any significant extent, as they are no longer available to the general public. New chemicals, mainly therapeutic drugs, are becoming available at some pace. And new medicines become new poisons. We are reminded of Alfred Swaine Taylor's statement: "A poison in a small dose is a medicine but a medicine in a large dose is a poison." Legalised killing is a growth area in terms of euthanasia and abortion, and more countries are considering lethal injection as a means of executing murderers. It is an interesting picture, full of hopes and fears.

Whatever developments take place, our fascination with poisoning as a means of ending human life will continue, especially where poison is used as a weapon of stealth.

Poisoners and Victims: A Summary of Poisoning Cases, Fatal and Non-fatal

Adams, John Bodkin. 1899–1983. Killed 160 plus patients using morphine and barbiturates His mother was said to be the holiest woman in Ireland. His father was a strict disciplinarian and a Plymouth Brethren preacher. Adams qualified to practice medicine but proved to be less than reliable, and at times incompetent. Throughout his medical career he was keen to receive money from his patients. In fact, he was described by one as a "real scrounger".

In 1957 he was brought for trial after investigations into around 400 wills that had been changed to make Adams the beneficiary instead of the deceased's family, as would have been expected. More than 160 of his patients had died suspiciously between 1946 and 1957. The wills were those of elderly female patients, many of whom appeared to have died from massive doses of morphine. He was convicted for fraud but not for murder, despite the fact that there was strong evidence of killings.

Allitt, Beverley Gail. Serial murderer. Born in 1968 she seems to have had a normal family background. She had two sisters and a brother. Her father worked in an off-licence and her mother was a school cleaner. Allitt developed worrying patterns of behaviour.

Poisons and Poisonings: Death by Stealth
By Tony Hargreaves

Published by the Royal Society of Chemistry, www.rsc.org

She would bandage herself to cover an alleged injury, but would not let others see the wound. It was an attention-seeking behaviour. She went to secondary school and was keen on babysitting. As an adolescent she became overweight and more attention-seeking, and showed aggressive tendencies. She often sought medical treatment and spent many hours in hospital when in fact she was physically in good health. There were instances of her self-harming. Doctors soon became aware of her and her efforts to fain illness or injury. In one instance, she faked appendicitis and had a healthy appendix removed. After the surgery she would interfere with the scar and prevent its healing.

Allitt went on to study at Grantham College and train as a nurse. Whilst working in a nursing home, Allitt showed some bizarre behaviour when she smeared faeces on the walls. During her period of training she was often absent, claiming one illness after another despite being physically well. Often she failed the exams. Her boyfriend said that she was deceitful, manipulative and aggressive. She claimed to be pregnant and to have been raped, both claims were false.

She found work as a junior staff nurse, on a six-month contract, on a paediatric ward at Grantham and Kesteven Hospital in 1991. The ward was then seriously under-staffed. There were only two nurses on the day shift and one at night. This staff shortage probably accounted for Allitt being able to continue with her attention-seeking behaviour, and allowed it to pass unnoticed and develop to violence against her young patients.

Allitt claimed her first victim on 21st February 1991. Seven-week-old Liam Taylor, had been admitted with a chest infection. Whilst in Allitt's care the baby suffered a respiratory emergency but recovered. She then volunteered for extra night duty so she could look after him. During the night Liam had another respiratory crisis but survived it. However, later, while alone with Allitt, his condition became worse, he turned pale, and developed red patches on his skin.

Allitt called the resuscitation team when Liam stopped breathing. It then became evident that the breathing monitor should have sounded the alarm but, curiously, it did not. Liam had a heart attack, suffered brain damage and was kept alive on a life-support machine. The baby died soon after, the cause of death was recorded as heart failure, but Allitt was not investigated.

Eleven-year-old Timothy Hardwick was admitted on 5th March 1991 after an epileptic fit. He was to become Allitt's second victim. While alone with the boy she called the resuscitation team who saw that he had no pulse and was turning blue. He died, but there was uncertainty as to the cause. As such it was regarded as resulting from epileptic fit.

Kayley Desmond, a one-year-old baby, was admitted to the ward with a chest infection and appeared to be recovering. However, five days after her admission Kayley had a cardiac arrest while she was being nursed by Allitt. Again Allitt called the resuscitation team, and they saved her. Kayley was sent to Nottingham hospital where specialists gave her a detailed examination and discovered a suspicious puncture hole under her armpit and a nearby air bubble. The conclusion was that this was due to an accidental injection, but no investigation was called for.

Allitt's next victim was five-month-old Paul Crampton. He was admitted to the ward on 20th March 1991 with a bronchial condition. Allitt, alone with the baby, called doctors when he appeared to be going into a coma. He was revived but the doctors were concerned as to why he had such a high insulin level. Paul was rushed by ambulance to Nottingham hospital with Allitt caring for him during the journey. Paul was again found to have high insulin levels. Despite having had been given an insulin overdose on three occasions, he survived the attacks.

The next day, five-year-old Bradley Gibson was brought to the hospital suffering from pneumonia. He had a heart attack but was saved by the resuscitation team. The doctors were surprised to find that Bradley had a high insulin level. Bradley, while in Allitt's care, had another heart attack during the night and was taken to Nottingham hospital and soon recovered. Even at this stage, no-one indicated suspicion relating to Allitt.

On 22nd March 1991, two-year-old Yik Hung Chan was in Allitt's care when he turned blue. A doctor gave him oxygen and soon he recovered. But later, he became blue again and was taken to Nottingham hospital.

The twins Katie and Becky Phillips were only two months old, but had been kept in hospital due to their premature birth. Becky fell ill with gastro-enteritis on 1st April 1991. Allitt was the nurse caring for her but two days later there was some alarm when Becky appeared to be cold and hypoglycemic. Later, all seemed well and the baby was allowed to go home. However, Becky went into convulsions during the night. She was seen by a doctor who thought the baby had colic. That same night Becky died, but a post mortem showed no clear cause of death.

Becky's sister, Katie, was then admitted to the hospital for observation as a precautionary measure and Allitt was responsible for her. Soon Katie had stopped breathing. The resuscitation team were called and Katie recovered until two days later when she had a repeat attack, which resulted in lung damage. She was taken to Nottingham hospital where she was found to have five broken ribs along with brain damage due to oxygen shortage. The baby survived but was left with cerebral palsy and damage to sight and hearing.

More incidents followed in which patients were attacked whilst in the care of Allitt. Suspicion was aroused and Allitt's killing career was brought to an end when fifteen-month-old Claire Peck had a cardiac arrest and died. Claire's body was exhumed and further tests revealed the presence of potassium and lignocaine in her system. Lignocaine is used during cardiac arrests, but would never be given to a baby. The police became involved, and their attention turned to previous deaths of children that were being cared for by Allitt in the two months prior to Claire's death. High levels of insulin were also found. It was also suspected that Allitt had smothered a child, had cut off the oxygen supply to another, and had tampered with life-support machines.

In May 1993 she was convicted and given 13 life sentences for murder and attempted murder. This included 4 counts of murder and 11 counts of injury and grievous bodily harm. She was thought to be suffering from Munchausen syndrome by proxy and was detained in Rampton Special Hospital. The poisoned children and dates of their deaths were: Liam Taylor, 21st February 1991; Timothy Hardwick, 5th March, 1991; Becky Phillips, 1st April 1991; and Claire Peck, 22nd April 1991.

Cotton, Mary Ann. Britain's first serial poisoner (Figure 1). She used arsenic and killed for financial gain or to remove one husband so as to begin a relationship with a new lover. It is estimated that she poisoned around 15 people, many of them children. Exact figures were impossible to establish as many people living in the poor circumstances that Cotton had to endure died of gastric fever, which had similar symptoms to chronic arsenic poisoning. Furthermore, she frequently moved from one area to another and she used a variety of names.

Figure 1 Mary Ann Cotton was a 19th Century serial poisoner. She used arsenic and killed for financial gain or to remove one husband so as to begin a relationship with a new lover. It is estimated that she poisoned around 15 people, many of them children. Exact figures were impossible to establish as many people living in the poor circumstances that Cotton had to endure died of gastric fever, which has similar symptoms to chronic arsenic poisoning.
Image courtesy of www.gutsandgore.com.

Cotton was hanged at Durham Prison in 1873 but swung for three minutes as the hangman William Calcroft had got the length of the rope wrong. We may wonder why, that after 40-odd years as hangman he managed to use the wrong length of rope for the drop. It left Cotton dangling as she slowly throttled to death. How did he get it wrong? There are reports of Calcroft making the same mistake repeatedly in earlier hangings. There are claims that Calcroft was incompetent.

Was this done deliberately to increase the suffering? Some said he was a callous showman. The latter is likely, for on several occasions when the criminal was left swinging and throttling, Calcroft would grab hold of the victim to bring about the final swing that broke the neck and ended it all. It was a show, for there were often crowds of many thousands, watching to see justice being done, or indulging a morbid curiosity.

The hangman is on stage, entertaining his audience. We would be justified in thinking that the executioner was no better than the criminal. In this context we might ask why a man should seek the job of hangman. Were these men psychologically damaged like many of those who committed murder?

> ***Nursery rhyme. Mary Ann Cotton.***
> *Mary Ann Cotton, she's dead and she's rotten.*
> *Lying in bed with her eyes wide open.*
> *Sing, sing, oh what shall I sing?*
> *Mary Ann Cotton, she's tied up with string.*
> *Where, where? Up in the air.*
> *Selling black puddings, a penny a pair.*
> *Mary Ann Cotton, she's dead and forgotten,*
> *Lying in bed with her bones all rotten.*
> *Sing, sing, what can I sing?*
> *Mary Ann Cotton, tied up with string.*

Ansell, Mary Ann. Used phosphorus for poisoning. Hanged by James Billington at St Albans Prison in 1899.

Armstrong, Herbert Rouse. Administered arsenic to poison his wife for financial gain. Hanged by John Ellis at Gloucester Prison in 1922.

Aum Shinrikyo religious group. Japan. Murdered using VX and Sarin chemical weapons in attack on Tokyo subway in 1995.

Bartlett, Adelaide. Chloroform found in her dead husband's stomach in 1885. Acquitted.

Bateman, Mary. The Yorkshire Witch. Selling poisonous medicine for financial gain. Hanged at York Castle by William Curry in 1809. Her skeleton and plaster cast death mask are sometimes on display at The Thackray Medical Museum in Leeds.

Barry, Mary Anne. Hanged by Robert Anderson at Gloucester Jail in 1873. She was the last woman to die by the short-drop method of hanging in Britain (Figure 2).

Figure 2 Many of those found guilty of murder by poisoning were hanged. In Britain, hanging as a means of lawful execution was used up until 1964. There are basically two types of hanging, and both use a rope noose placed around the neck of the convicted person. In the 'short drop' method the object supporting the person is pulled from beneath them causing their body to drop a short distance during which the noose tightens and death by strangulation occurs. This could take up to 15 minutes. It was regarded as inhuman and so the procedure was improved to cause a rapid death. This led to the British 'long drop' method in which the person dropped a longer distance before the noose tightened. In a fraction of a second the neck was broken and death resulted.
© Shutterstock.

Berry, Elizabeth. Administered creosote as poison, hanged at Walton Jail by James Berry in 1887.

Biggadyke, Priscilla. Carried out poisoning with arsenic. Hanged by Thomas Askern at Lincoln Prison in 1868.

Blandy, Mary. Used arsenic to poison her father. Hanged 1752.

Brahe, Tycho. Astronomer, accidental poisoning with mercury whilst doing experiments on elixir. Died 1601. Presence of mercury detected in his hair.

Britland, Mary Ann. Poisoned her lover's wife, her own husband and her own daughter. Hanged by James Berry at Strangeways Manchester Prison in 1886.

Bryant, Charlotte. Poisoned her husband with arsenic. Hanged by Thomas Pierrepoint at Exeter Prison in 1936.

Buchanan, Dr Robert. Morphine poisoning for financial gain. Executed by electric chair in 1895 at Sing Sing Prison, America.

Chapman, George. Used antimony to kill three or more between the years 1897–1902. Hanged at Wandsworth Prison in 1903.

Christie, John Reginald. Carbon monoxide poisoning and strangulation, during the years 1943–1953 He killed eight plus. Hanged at Pentonville Prison in 1953.

Chua, Victorino. Injected insulin into saline bags and ampoules to kill patients. Given a life sentence in 2012 with a minimum of 35 years. Two murders and 31 other charges of poisoning patients at Stepping Hill Hospital, Stockport.

Clarke, Frances. In 1817 she killed her infant son by making him swallow sulphuric acid (oil of vitriol) purchased from a local shop. Sentenced to death but then given a pardon.

Clements, Dr Robert George. Suspected of poisoning wives with morphine but committed suicide in 1947 before he could be brought to trial. The motive was likely to have been financial gain.

Cream, Dr Thomas Neill. The Lambeth Poisoner. A psychopath who was motivated by sadism and hatred. Used strychnine to kill several prostitutes. Possibly also a chloroform poisoner.

Murdered 8 women and 1 man between the years 1881–1882. Executed by James Billington at Newgate Prison, London in 1892.

Born in Glasgow in 1850. Studied medicine at McGill University in Canada. Graduated 1876. Moved to London and began training at St Thomas's Hospital to be a surgeon but failed the exam. Later he was accepted by the Royal College of Physicians and Surgeons in Edinburgh.

Whilst living in London he developed a hatred for prostitutes in the poverty stricken East End. In 1878 he returned to Canada and set up a practice as a physician. He moved to America and set up a practice near to a red light district. Performed illegal abortions. His hatred of prostitutes increased despite having sex with them. At this time he was taking pills of strychnine, morphine and cocaine for their aphrodisiac effects.

Although not all cases were proven, he was implicated in the following murders: Kate Gardener, chloroform, Canada; Flora his wife, strychnine, Canada (he also aborted their baby); Mary Anne Faulkner, killed during abortion, America; Ellen Stack, strychnine, America; and Daniel Stott, his only male murder, strychnine, America. After killing Stott, he was convicted and sent to Joliet Prison, Chicago. 1881. Released in 1891.

He returned to London and was implicated in further murders: Ellen Donworth, strychnine, London; Matilda Clover, death was recorded as due to alcohol but likely to be strychnine as she'd taken pills given to her by Cream, London; Alice Marsh, strychnine, London, 1892; and Emma Shrivell, strychnine, London, 1892.

Crippen, Dr Hawley Harvey. Used hyoscine to kill his wife. Executed at Pentonville Prison by John Ellis in 1910.

Croydon poisonings. 1928–1929. Many people died but the culprit was never found. Arsenic was found in the exhumed body of one. Some others showed symptoms of arsenic poisoning.

Deshayes, Catherine. Poisoned several people. Convicted of witchcraft and burned in public in Paris, 1680.

Devereux, Arthur. Used morphine as a poison. Hanged by Henry Pierrepoint at Pentonville Prison in 1905.

Dove, William. Used strychnine as a poison to kill his wife. Executed in 1856, watched by over 10 000 spectators.

Dubcek, Alexander. Survived poisoning attempt with strontium-90.

Eccles, Betty. Poisoned her stepson with arsenic. Executed at Liverpool Prison in 1843.

Edmunds, Christina. Used strychnine as a poison in 1971 with passion being the motive. Sentenced to death but commuted to life in Broadmoor.

Emery, James. Used arsenic as a poison. Hanged in 1831.

Flanagan, Catherine. Poisoned Margaret Higgins' husband using arsenic for financial gain from life insurance. Other killings likely. Hanged along with Margaret Higgins at Kirkdale Prison Liverpool by Bartholomew Binns in 1884.

Foster, Catherine. Used arsenic as a poison. Hanged in 1847.

Galley, Ada Charlotte. Used chlorodyne as a poison, hanged with accomplice Sarah Walters by Henry Pierrepoint at Holloway Prison in 1903.

Gilbert, Kristen Heather. Had a history of violent threats and faked suicides. Two attempted murders and three murders. Adrenaline injections into therapy bags caused cardiac arrest. Life sentence, Massachusetts, America.

Hahn, Anna Marie. Used arsenic as a poison for financial gain. Killed five people in America. Executed by electric chair at Ohio Penitentiary in 1938.

Hamilton, Susan. Put sodium chloride in her daughter's feeding tube.

Harvey, Donald. He was an American serial poisoner who called himself the Angel of Death and claimed to have murdered 87 but the police estimates show it is somewhere between 37 and 57. The murders took place between 1970 and 1987. He worked as an orderly in a hospital where he murdered patients who were terminally ill. Various poisons were administered, for example, arsenic, cyanide, insulin and morphine. Harvey's evil ways were not confined to poisons as on one occasion he pushed a coat hanger into a patient's catheter, which caused abdominal puncture and peritonitis. At school he got on well with his teachers and obtained good grades but pupils regarded him as being a loner. It is alleged that he was sexually abused by his uncle and a male neighbour.

Hazel, Nancy. Also known as Nannie Doss. American serial killer. Killed her own mother, four husbands, one mother-in-law, two children, two sisters and a grandson.

Higgins, Margaret. Hanged with sister Catherine Flanagan for poisoning Flanagan's husband. Other killings likely.

Hitler, Adolf. Hitler was a mass poisoner, a serial poisoner and possibly committed suicide with poison. Born 1889, Hitler had an unhappy childhood. His father, who died when Hitler was 16, was a customs official and was violent to the young boy's mother and to the boy himself. Hitler was deeply attached to his mother but rebelled against his father. The boy was good at school and he enjoyed painting, which gave him ambitions to pursue this as a career but he never made the grade.

In 1913 he moved to Munich to avoid military service in his home country, Austria. This led to his arrest by the Austrian army but they found him to be unfit for service and released him. However, at the outbreak of World War I he joined the Bavarian army and was awarded the Iron Cross, but was not promoted to full corporal. When the war ended, Hitler remained in the army and joined the German Workers' Party (later to become the National Socialist German Workers Party). This nationalist organisation appealed to him as it held antisemitic and anticommunist views.

Hitler was discharged from the army in 1920 and became more involved with the party. During this time he became skilled at oratory and attracted large audiences. The rise of the Nazi party took Hitler to being the leader, der Fuhrer. In 1935 Hitler increased the size of the German army to 600 000, expanded the navy and introduced the Luftwaffe. He also brought in new laws regarding Germany's Jewish population.

Hitler's systematic killing of the Jewish people began in 1939. From then until 1945, he killed between 11 and 14 million people, which included 6 million Jews. The main method of killing was by using poisonous gases. This began with carbon monoxide but then prussic acid replaced it.

The details surrounding Hitler's supposed suicide are unclear and theories abound. Were there plans for him to escape and start the Fourth Reich in South America? Or did he commit suicide, as is generally believed, and was his body burnt with petrol? Where is the jaw bone purported to match his dental record? One theory is that he didn't commit suicide but escaped through a tunnel joining to the underground train network. The theory suggests he was then flown in Junkers 52 aircraft (short take-off and landing) taking off from a makeshift runway. It is claimed that the pilot flew Hitler and his wife to Tonda Denmark and then to Argentina.

Jones, Genene. Nurse, serial poisoner.

Jonestown massacre: This was the Peoples Temple Agricultural Project established by American Jim Jones and moved to Guyana in 1977. Over 900 Americans died there by cyanide poisoning in a 'revolutionary suicide'. It is arguable as to whether this was mass suicide or mass murder.

Khalashikar, Viktor. Russian ex-KGB colonel and wife. Survived mercury poisoning attempt in 2010.

King George III of England. A sample of his hair was analysed in 2004 and contained high levels of arsenic. It is suggested that the source of the arsenic was the antimony medication that he was taking. At the time, the extraction of antimony to prepare the medicine was not well understood and the antimony could have been contaminated with large amounts of arsenic.

Kuklinski, John. Cyanide poisoning for financial gain. Charged in 1988 and given two life sentences.

Lefley, Mary. Poisoned her husband with arsenic. Hanged by James Berry at Lincoln Prison in 1884.

Lamson, Dr George Henry. Financial gain, aconitine, poisoned wife's brother, executed by William Marwood at Wandsworth Prison in 1882.

Lipski, Israel. Used nitric acid as a poison. Hanged at Newgate Prison by James Berry in 1887.

Major, Ethel Lillie. Poisoned her husband with strychnine. Hanged by Thomas Pierrepoint at Hull Prison in 1934.

Markov, Georgi. Victim of assassination. Poisoned with ricin.

Maybrick, Florence Elizabeth. Poisoned her husband with arsenic extracted from fly papers. Imprisoned in 1889 but released in 1904.

Merrifield, Louisa May. The Blackpool Poisoner. Using Rodine, a phosphorus-based rat poison, killed her employer for financial gain. Hanged by Albert Pierrepoint at Strangeways Prison Manchester in 1953.

Monroe, Marilyn. A most complex case with many unanswered questions. She was probably a victim of medical error.

Norris, Colin. Pathological hatred of old people. He worked as a nurse in Leeds where he injected high doses of insulin into his patients. Seventy-two cases were investigated by the police. He was convicted of the murder of four patients aged 79 to 88 and sentenced to life imprisonment in 2008 to serve a minimum of 30 years. As of March 2016, this case is being reviewed by the Criminal Cases Review Commission because of new scientific evidence.

Overton, Hannah. Used salt to poison a foster child.

Palmer, Dr William. Administered cyanide, antimony and strychnine. Killed three plus. Hanged at Stafford Prison in 1856
Patrick, Jacqueline. Given a 15-year sentence for spiking cherry Lambrini with antifreeze to kill her husband.

Pearson, Elizabeth. Killed her uncle with rat poison. Hanged by William Marwood at Durham Castle in 1875.

Pritchard, Dr Edward William. Used antimony to killed his wife and mother-in-law. Hanged by William Calcraft at Glasgow Prison in 1865 in front of a crowd of 100 000.

Seddon, Frederick. Used arsenic as a poison. Killed one person for financial gain. Executed at Pentonville Prison by Thomas Pierrepoint in 1912.

Sevango, Dr Michael. Potassium chloride injection.

Shipman, Dr Harold Frederick. Administered diamorphine, during the years 1975–1998. His serial poisonings killed an estimated 300 patients. Sentenced to life imprisonment. Shipman had a domineering mother who instilled into him a sense of superiority. He developed an arrogance that made it difficult for him to make friends and led to him becoming an isolated adolescent. She had lung cancer and he took charge of her care. During this time, it seems he was fascinated by the powerful positive effect that morphine had.

When his mother died, Shipman became determined to study medicine and become a doctor. He got a place at medical school after failing the entrance exam first time round. At the age of 19 he married a girl of 17 who was five-months' pregnant. After his medical studies he found a job in a small practice and allegedly became addicted to pethidine. His need for a constant supply of the drug caused him to forge prescriptions for large amounts, which led to him being caught and dismissed.

After leaving his job he went on a drug rehabilitation programme. For his forgery of prescriptions, he was made to pay a small fine and was also given a conviction for forgery. However, he was still allowed to continue as a doctor and once again obtained a post in a general practice where he was hard-working and respected, but tended towards arrogance when dealing with younger staff.

Over a period of years, he had taken to poisoning a series of his older patients by overdosing them with diamorphine. It was years before the authorities noticed the unusually high number of deaths, but eventually the police charged him with murder when vital evidence relating to a will that he had forged came to light.

At his trial he did himself no favours by continuing with his arrogance when questioned by the police. Even at his trial he was arrogant and changed his story several times. He hanged himself in Wakefield Prison in 2004.

Swanenbyrg, Maria. Holland. Thought to have poisoned more than 80 people.

Tofana, Giulia. Arsenic, tortured and strangled in prison, 1709.

Towell, John. Cyanide poisoning in 1845 to cover up an illicit affair. Hanged in front of a crowd of 10 000.

Taylor, Louise Jane. Poisoned using sugar of lead for money and sadistic pleasure. Hanged by William Marwood at Maidstone Prison in 1883.

Turing, Alan. Committed suicide by eating an apple containing cyanide.

Waddingham, Dorothea Nancy. Referred to as a 'nurse'. Set up a nursing home in Nottingham to care for the elderly and infirm. Used morphine to poison two women, mother and daughter, for financial gain. Hanged by Thomas and Albert Pierrepoint at Winson Green Prison Birmingham in 1936.

Wainright, Thomas Griffiths. Used strychnine to kill four plus for financial gain. Transported in 1820.

Watson, Lionel Rupert Nathan. Used sodium cyanide as a poison. Executed at Pentonville Prison by Thomas Pierrepoint in 1943.

Wilson, Catherine. Nurse. Poisoned victims with colchicum for financial gain between the years 1853–1862. Hanged at Newgate Jail in 1862 in front of a crowd of 20 000.

Wilson, Mary Elizabeth. Used phosphorus as a poison between the years 1955–1957 to kill five plus. Sentenced to life imprisonment.

Woolcock, Elizabeth. Mercury poisoning, hanged 1874. Some unanswered questions in this case.

Young, Graham. Used thallium and other poisons during the years 1962–1971. He killed at least three and non-fatally poisoned around 70 people as he experimented with substances and dosages. At the age of only 14 he was sent to Broadmoor high-security hospital for a stipulated period of 15 years. He was released after eight years on the recommendation of a psychiatrist who thought Young was a reformed person and would not do any more experiments.

After his release he obtained a job in a laboratory where highly toxic thallium compounds were used. He had chosen the kind of occupation that would enable him to continue his poisoning experiments and hoodwinked the authorities. He was careful, creative, cunning and sneaky. Furthermore the disruptive experiences in his childhood could well have had a negative effect upon his developing psychology.

He was arrested and sent to Parkhurst Prison where he died in 1990. During his stay there he made friends with Ian Brady, the moors murderer, and was delighted to find that a waxwork model of him was on display at Madame Tussauds Chamber of Horrors. In the display, Young was placed beside the murderer Dr Crippen who was one of Young's heroes.

In his serial poisonings, Young aimed to poison his victim as a form of punishment or to kill them. There seemed to be two separate motives. His behaviour profile showed an obsession with poisons. He experimented with different substances at various dose levels and meticulously recorded the results. Young's victims were mere guinea pigs.

Yushchenko, Viktor. Ukrainian politician. Survived dioxin poisoning attempt in 2004.

Glossary

There is often confusion over the names of chemicals. For example, there are two distinctly different chemicals that are sometimes called 'mercury chloride'. There is mercurous chloride and mercuric chloride; they have different chemical formulae, different physical properties, different reactions, and different toxicities. With the modern naming system we call mercurous chloride mercury(I) chloride and mercuric chloride is now mercury(II) chloride.

But these two chemicals also have common names. In the literature on poisons we often find the older names for these. Thus, we have 'calomel' for mercurous chloride and 'corrosive sublimate' for mercuric chloride. The distinction is important as there's a huge difference in how poisonous they are. Calomel was put into teething powder for babies, and corrosive sublimate is what the mad hatters suffered brain damage from.

It should also be noted that the name, or parts of it, should not be taken at face value. The chemical called 'theobromine' does not contain bromine, and the word 'mercaptan' is not meant to imply a connection with mercury.

There have been a few attempts to rationalise the naming of chemicals. An early example relates to the chemical called 'dephlogisticated muriatic acid gas'. It is a rational name and conveys meaning so long as you know what muriatic acid is, and

Poisons and Poisonings: Death by Stealth
By Tony Hargreaves

Published by the Royal Society of Chemistry, www.rsc.org

you are familiar with the phlogiston theory. That name, which I think has a certain charm, tells us that the gas is the one we get from oxidising hydrogen chloride. In the gas attacks at Ypres it was called Bertholite. Today we call it 'chlorine'.

When considering the finer details of poisons and poisoning cases it is essential to be precise. To overcome this, and to be sure of an identity of a chemical compound or a chemical element, a universal identifier is used. It is the Chemical Abstracts Registry Number (CAS Registry Number). It applies only to pure substances, which are defined by their chemical formula. It does not apply to mixtures of chemicals which, by their nature, have no precise formula and so there may be an infinite number. In the CAS registry we find our mercury(I) chloride is 10112-91-1 and the mercury(II) chloride is 7487-94-7.

New synthetic chemicals are being invented each week, and hitherto unknown natural chemicals are being discovered. When these are purified and defined by a formula they wait to be allocated their unique CAS number.

As of February 2016 there were 106 million chemicals listed and new ones join the queue each day. Just how many of these are poisonous? Clearly contemporary technology is bringing us masses of new poisons. Of course, the intent is to offer us new medicines. But where there's a medicine there's a poison. Already we have thousands more chemicals than could have been imagined a century ago.

abortifacients: Mifepristone is a synthetic drug used medically for abortion. It works by blocking the action of the progesterone. Sometimes it is used in combination with gemeprost, which causes powerful contractions of the uterus. Pulegone and menthone are natural abortifacients found in pennyroyal, (*Mentha pulegium*), which is a member of mint family.

aconine: see aconite.

aconite: The dried root of wolfsbane (*Aconitum napellus*). It contains three analgesic compounds: aconine; aconitine; and picraconitine. Aconite was used as the painkilling agent in toothache tincture. No longer used due to its poisonous properties.

aconitine: see aconite.

adrenaline: Epinephrine (America). Hormone naturally produced in humans. Acts on heart rate and breathing rate to prepare the body for 'flight or fight'. May be administered by injection in medical emergency such as cardiac arrest. Has been used in overdose in murders.

alkaloids: Natural compounds mainly of plant origin, but also a few from animals. They all contain nitrogen, which makes them react as bases or alkalis, hence the name 'alkaloid'. Name of alkaloid ends with –ine and indicates the source. See individual entries.

amanitin: Poisonous compound found in certain mushrooms of the group *Aminata phalloides.*

amygdalin: A cyanogenic compound found in bitter almonds, peaches and apricots.

antimony: (a) Stibine is antimony hydride and about as toxic as arsine gas. (b) Tartar emetic was based upon antimonial potassium tartrate but is no longer used due to its poisonous nature.

arsenic: Scheele's green, copper arsenite, a vivid green pigment once used in oil paints and wallpapers. It releases extremely poisonous arsine gas (arsenic hydride) and cacodyl gas (tetramethyldiarsine), in damp conditions. White arsenic has been used in many poisonings.

arsine: see arsenic.

atropine: An alkaloid found in the plant deadly nightshade, *Atropa belladonna*. Also present in henbane (*Hyoscyamos niger*) and mandrake (*Mandragora officinarum*). The latter is referred to in Shakespeare's *Othello* III 334. In small doses atropine behaves as a muscle relaxant and reduces secretions such as salivation.

back-calculation: In drink-driving cases a blood sample may be taken some time after the incident. This calculation enables the level of alcohol present to be assessed and related to when the incident occurred. $C_0 = C_t + t\beta$ where C_0 is the blood alcohol concentration at some initial time, C_t is the blood alcohol concentration at a later time, t is the time that has elapsed between the initial and later times, and β is the rate of elimination of the alcohol from the blood.

barbiturates: Synthetic compounds derived from barbituric acid, and useful as prescription sedatives. Little used these days as they are habit-forming and can lead to tolerance of large doses which, in combination with alcohol, can be fatal. They act as central nervous system depressants. Examples: amobarbital; butobarbital; phenobarbital; and pentobarbital (Nembutal).

batrachotoxin: Highly toxic alkaloid found in the skin of dart frogs, which have glands to produce the chemical.

belladonna toxins: A mixture of poisonous alkaloids from the plant deadly nightshade (*Atropa belladonna*). Examples: atropine; hyoscine (scopolamine); and hyoscyamine.

bio-accumulation: Poisons that are organic compounds are generally broken down in the body by the liver and, depending on whether the person survives or dies, are excreted in the urine or remain in the dead body. However, some poisons, such as the heavy metals, are elements and cannot be broken down. They can, if the person survives the poisoning, be isolated from the body's chemistry by being stored in the bones, teeth, hair and nails. In effect, the heavy metal is being taken away from where it can do harm.

As such, it is excreted but it is not eliminated from the body. Of course, a single fatal dose of a heavy metal would poison the victim before any of the metal had found its way to the bones, hair, *etc.* The bio-accumulation of heavy metals is usually seen with long-term exposure to the poison and, as such, the forensic analysis would focus on, say, a sample of the hair.

Examples: (a) Lead bio-accumulation is found in many people in the industrialised World due to long-term exposure to lead fumes and dust, the main culprit being the lead tetraethyl added to petrol. There must many people alive today who have lead poisoning due to the chronic effect of lead on the central nervous system. (b) In nature, bio-accumulation occurs naturally, such as when oysters absorb arsenic from the sea water and store it in their flesh.

bio-available: This refers to the proportion of poison that produces a physiological effect. For example, if heroin is swallowed it is not completely bio-available, but if injected it is 100% bio-available.

botulinum toxin: A powerful nerve toxin produced by the bacterium *Clostridium botulinum.* The toxin is responsible for one of the more serious forms of food poisoning, in which the central nervous system is attacked. Death occurs due to heart and lung failure. The bacterium is usually found in foods that have not been correctly preserved, such as canned meats that have not been given sufficient heat treatment.

bromodiolone: Rat poison based upon coumarin. Attacks Vitamin K function, preventing its blood-clotting action. Subsequent internal haemorrhage is fatal. Introduced in the 1980s to replace other rat poisons that had become ineffective as the rodents had developed a resistance.

bufotoxin: A toxic secretion from many species of toad. Some of these toxins contain bufalin, adrenaline and serotonin.

cacodyl: see arsenic.

cadaverine: see ptomaines.

caffeine: An alkaloid found tea and coffee. One of the less poisonous alkaloids. It behaves as a stimulant of the central nervous system.

cantharidin: Vesicants secreted by the blister beetle, Spanish fly. Its chemical structure is basically that of the terpenes. Spanish fly is the blister beetle *Lytta vesicatoria*, which produces the poison cantharidin when threatened.

carbolic acid: Phenol. A poisonous antiseptic manufactured from petroleum. A few cases of people swallowing phenol solutions and creosote have been reported. Cough mixtures once contained small amounts of phenol and related compounds known as 'cresols'. If undiluted, the carbolic acid burns the oesophagus and can lead to choking and asphyxiation.

carbon monoxide: Poisonous gas that works by blocking the oxygen-carrying property of the red blood cells. Occurs in vehicle exhaust gases, and can be formed when gas appliances do not burn efficiently. It was the poisonous component in the coal gas at one time supplied to consumers. This domestic gas used in the past contained about 4% carbon monoxide and caused many poisonings, particularly suicide cases.

ceruse: see lead.

chloral hydrate: A sedative and hypnotic substance taken by mouth. It was introduced into medicine in Britain in the 1870s as a sedative that could be taken by mouth as a solution. Chloral hydrate crystals were readily available without having to sign the poison book. Sometimes called 'knockout drops'.

The drink known as a Mickey Finn was a solution of chloral hydrate in ethanol. Traces of chloral, along with chloroform, are to be found in the water supply where chlorine has been added to disinfect it. This is due to chlorine reacting with humic acids dissolved in the water, especially in peat areas. Taking chloral on a long-term basis can lead to addiction, kidney, heart and liver failure. Chloral can be supplied in suppository form. Chloral was used in Jonestown in the extended care unit for those who tried to escape. Some were hung upside down.

chlorodyne: A popular medicine in 19th Century Britain. It contained alcohol, opium, tincture of cannabis and chloroform. It was highly effective for relief of pain, as a sedative, and for curing diarrhoea.

chlorinated hydrocarbons: Many are volatile compounds that have an effect on the central nervous system and can cause poisoning by inhalation and skin contact. Also cause damage by dissolving body fats. Examples: (a) chloroform, trichloromethane, formerly used as an anaesthetic; (b) trichloroethylene, formerly used as an anaesthetic; (c) halothane, a modern inhalation anaesthetic; (d) DDT, a well-known synthetic pesticide that caused severe environmental poisoning; and (e) TCDD, tetrachlorodibenzodioxin, or 'dioxin' is highly toxic and causes chloracne.

cicutoxin: Toxic alkaloid found in water hemlock, cowbane (*Cicuta maculate*).

cocaine: An alkaloid that has a powerful effect on the central nervous system. Was used as an anaesthetic but now discontinued. Used as a recreational drug but many fatalities due to cocaine poisoning occur.

codeine: An alkaloid found in opium and similar to morphine. As codeine phosphate it is used as a painkiller, but can be habit-forming.

coniine: Toxic alkaloid found in poison hemlock (*Conium maculatum*). Thought to be the poison that killed Socrates. Hemlock also contains methylconiine and conhydrine.

core temperature: After death the body temperature gradually changes to the temperature of the environment. Usually this means cooling from 37 °C. If the temperature of the environment is about 18 °C, the core temperatures would typically be as follows: 6 hours = 32 °C; 10 hours = 28 °C; and 15 hours = 25 °C. At about 40 hours the exothermic decomposition reactions are firmly established and the temperature increases a little.

curare: An arrow poison extracted from the bark of various trees such as *Chondodrendron tomentosum.* It works as a muscle relaxant and neurotoxin. The active component is tubocurarine.

cyanide: There are many cyanide chemicals, but it is the water-soluble compounds that are deadly poisonous. This is because these compounds release cyanide ions in solution. Examples: (a) Hydrogen cyanide, also known as 'prussic acid' is treacherously poisonous because it can be inhaled whereupon it dissolves the moist membranes of the lung to form hydrocyanic acid. The gas was used in the Nazi gas chambers of World War II. (b) Potassium cyanide and sodium cyanide. Water-soluble and deadly poisonous. (c) Cyanogenic glycosides contain cyanide bonded to a sugar. When ingested they are broken down by digestive enzymes that cause the deadly poisonous cyanide ion to be released. Examples include amygdalin, which is found in bitter almond kernels, and prunasin, which occurs in cherries and peaches. Marzipan contains a significant amount of cyanogenic glycoside and could be poisonous if consumed in large quantities. Cassava is a common food plant in Africa. As it contains cyanogenic glycosides, it must be finely ground and thoroughly washed in running water to remove the poison.

cytisine: Toxic alkaloid found in lalburnum (*Laburnum anagyroides*).

death, cause of: This is the medical cause of death as stated by a doctor who has examined the body. It is written on the death certificate. Example: carbon monoxide poisoning. See also manner of death.

death, manner of: Not always recorded on the death certificate. Example: suicide.

diamorphine: Diacetylmorphine, heroin. Opiate drug made from morphine.

environmental poisons: Currently the main poisons from human activities that are polluting the oceans, the ground and the atmosphere and creating a potential threat to living systems are as follows: (a) soot, ash, and dust from burning; (b) arsenic and other heavy metals from metal smelting and coal burning; (c) pesticides and fertilizers from agricultural run-off; (d) methane from sheep and cows; (e) detergents, antibiotics, and microplastics in rivers and oceans; (f) PCBs and PAHs in contaminated ground; (g) dioxins from burning PVC; and (h) pathogens in raw sewage.

epibatidine: A poisonous alkaloid secreted by certain frogs. Can cause poisoning by skin absorption.

execution: see hanging.

ergotamine: Poison produced by ergot fungus (*Claviceps purpura*).

fatty acids: Carboxylic acids. A series of many compounds, a few of which are involved in the chemistry of the human body, especially that of body fat. During decomposition of a corpse, a stage may be reached in which butyric fermentation takes place with the release of volatile butanoic acid, which smells like cheese.

gemeprost: see abortifacients.

grayanotoxins: Poisonous compounds present in the honey from nectar of rhododendron flowers. Eighteen such compounds have been isolated from this source.

half-life: Time taken for the concentration of a substance in the body to fall to half of its initial concentration. Examples relating to poisons in the human body: brodifacoum rodenticide, 20–130 days; mercury (mercuric) chloride, 60 days; DDT pesticide, 3.7 months; and lead in bone, 25 years.

hanging: Where execution is by hanging there are two types, the 'short drop' and the 'long drop'. The former caused death by strangulation (garrotting, throttling, choking to death) whereas the latter was by means of breaking the neck. With strangulation, the prisoner could swing in agony for up to 10 minutes. Breaking the neck meant instant death. But not all long drops succeeded in breaking the prisoner's neck. When such failure occurred the hangman was known to jump into the drop pit onto the prisoner to add the extra force to break the prisoner's neck. The body hangs for the customary hour.

heroin: see diacetylmorphine.

hydrocyanic acid: see cyanide.

hydrogen sulphide: Sulphuretted hydrogen. A deadly poisonous gas. Smells of rotten eggs when at low concentrations but, as the concentration increases, the smell is no longer detectable as the gas desensitises the olfactory nerves.

hyoscine: Scopolamine. An alkaloid found in deadly nightshade, henbane and mandrake. Not to be confused with hyoscyamine, a different compound from the same source.

hyoscyamine: A poisonous alkaloid found in deadly nightshade.

indoor air quality: IAQ. The air indoors contains many unnatural vapours, gases and dusts. Many are irritants; some are poisons. Examples: (a) an office photocopier releases ozone form its ultraviolet light source; (b) out-gassing of freshly painted surfaces adds VOCs to the air; and (c) polymers release plasticizer fumes. In industrial processes the many chemicals used may degrade the workplace IAQ.

indicator species: Animals and plants can be highly sensitive to poisonous substances in the air and may be used as indicators. Examples: (a) canaries in coalmines, their behaviour can indicate poisonous gases in the air; and (b) lichen growing on rocks indicates the poisonous pollutants in the atmosphere.

lead: A soft grey elemental metal. Ancient Romans made sugar of lead to sweeten wine. Other examples: (a) lead acetate, sugar of lead, soluble; (b) red lead, a type of lead oxide formerly used as a food colouring; (c) ceruse, white lead, obtained from the mineral cerrusite; and (d) organic lead compounds, volatile and bioavailable through inhalation. Lead tetraethyl is the best known example of an organic lead compound. It was formerly put into petrol to improve combustion properties, but caused widespread lead poisoning by damage to the central nervous system.

lethal dose (LD_{50}): This represents the amount of a substance in mg kg^{-1} of body weight that will kill 50% of a species, usually rat, when administered orally. Examples: thallium sulphate = 25; arsenic trioxide = 15; potassium cyanide = 10; strychnine sulphate = 5; brodifacoum = 0.27; and ricin = 0.02 by injection.

mass poisons: Chemical warfare poisons. Examples: (a) Mustard gas, sulphur mustard, dichlorodiethylsulphide, colourless oily liquid, faint garlic-like odour, boiling point 216 °C, which makes it fairly involatile, persistent; (b) Sarin, nerve gas, the isopropyl ester of methylphosphonofluoridic acid; (c) Tabun, nerve gas, the ethyl ester of dimethylphosphoroamidocyanidic acid; (d) Soman, the methylpropyl ester of methylphosphonofluoridic acid; (e) VX, liquid nerve agent, persistent as it evaporates slowly, the ethyl ester of methylphosphonothoic acid; (f) Phosgene, carbonyl chloride, carbon monoxide and chlorine in one molecule; and (e) Riot control agent, CS gas, chlorobenzal malonitrile.

measurements: 1 grain = 0.0648 gram; 1 ounce = 28.3 gram; and 1 pound = 454 gram.

menthone: see abortifacients.

mercury: Also called 'quicksilver'. The only metal to be a liquid at room temperature. It releases vapour that is toxic by inhalation and forms many compounds that are toxic when ingested or absorbed through the skin. Poisoning symptoms: (a) Acute. Soluble salts have a violent corrosive effect on the skin and mucous membranes, severe nausea, vomiting, abdominal pain, bloody diarrhoea, kidney damage and death within 10 days. (b) Chronic. Inflammation of mouth and gums, excessive salivation, loosening of teeth, kidney damage, muscle tremors, jerky gait, spasms of extremities, personality changes, depression, irritability, nervousness. Some organic mercury chemicals can be absorbed by the skin. The liquid metal releases poisonous vapour and inhalation results in it becoming bio-available and behaving as a poison. Examples: mercuric nitrate; corrosive sublimate, mercuric chloride; organic mercury compounds, such as methyl mercury are readily absorbed through the skin and by inhalation, highly poisonous; mercuric cyanide, highly poisonous due to double action of mercury and cyanide; mercury compounds formerly used as medicines; and cinnabar, mercuric sulphide.

mifepristone: see abortifacients.

mineral acids: Sulphuric, nitric, hydrochloric, phosphoric. If swallowed have a corrosive effect on mouth and oesophagus, but cause instant vomiting so they are unlikely to be retained. Damage to living tissue occurs due to high level of acidity and not to a toxic effect. Not significant in terms of poisonings. Most cases due to accidental swallowing.

morphine: Also known as 'morphia' and 'morphium'. An opiate drug that is highly addictive.

muscarine: Deadly poisonous compound found in red-capped toadstools.

myristin: Nutmeg contains around 1.5% of this chemical, which breaks down in the liver to form MMDA, an amphetamine-class psychedelic chemical.

neurine: see ptomaines.

neurotoxin: Any poison that attacks nerve cells of the central nervous system (brain and spinal cord) and peripheral nervous system (all remaining nervous system).

nembutal: see barbiturates.

nicotine: Poisonous alkaloid from dried leaves of tobacco plant *Nicotiana tabacum.* Psychoactive component in cigarettes. Electronic cigarettes use solutions of nicotine of about 1.5%.

nitrosamines: Occur in latex products but not a poison risk in condoms. In foods nitrosamines are produced by nitrites reacting with amines, which occur in proteins. Produced in highly acidic conditions such as the stomach. Found in fish and meat products preserved with nitrites. Traces exist in beer. Some nitrosamines are known to be carcinogenic.

occupational poisoning: Many cases in the past where employees were exposed to poisonous substances. These days, less of a problem in the developed World due to regulation, risk assessment and workplace monitoring. With an understanding of poisoning it has been possible to draw up guidelines for good practice based upon maximum exposure levels for each hazardous substance. However, occupational poisoning kills many in less developed countries. Examples of past cases: (a) felt makers in the hat industry suffered mercury poisoning, mad-hatters disease; (b) London match girls suffered from phosphorus poisoning *via* a disease known as 'phossy jaw'; (c) radium poisonings to women painting luminous dials for clocks; and (d) munitions workers poisoned through inhalation of fumes from nitro compounds.

opium: A gum containing a mixture of alkaloids. Found in unripe capsules of the poppy (*Papaver somniferum*).

oxalic acid: Ethanedioic acid. Poisonous organic compound found in the leaves of the rhubarb plant (*Rheum rhaponticum*). Destroys kidney function by formation of calcium oxalate crystals, which block the capillaries. Oxalic acid is formed by oxidation when ethylene glycol is ingested.

ozone: Low-level ozone caused by vehicles emitting hydrocarbons in their exhaust gases is a recent concern, as it affects people exposed to city traffic fumes. The gas causes severe irritation of the respiratory tract and eyes.

pancuronium bromide: Used in legal killing of people during euthanasia or execution. Trade name is Pavulon.

paraquat: Dimethyl pyridinium (para quaternary), weedkiller, when swallowed has disastrous effect upon lungs. Paraquat poisoning is nearly always fatal.

pesticide: Natural or man-made poison for killing different living species, such as fungicide, bactericide, herbicide, insecticide, molluscicide, and rodenticide. Examples: (a) the herbicide 2,4-D acts upon the plant to destroy the function of the essential enzymes, causing it to die; (b) formaldehyde is the molluscicide used in slug pellets, but will also kill birds, mice and other animals that swallow the pellets; and (c) organophosphate compounds were investigated as pesticides in the 1930s. In Nazi Germany this research led the way to developing chemical warfare weapons.

phosphorus: Many forms of this non-metallic element exist with white phosphorus being the most reactive and spontaneously flammable, burning to form white fumes of phosphorus oxide. Phosphorus paste was used as a rat poison. Phosphorus is used in incendiary devices in warfare and is the basis of some chemical warfare agents (organophosphorus) and modern pesticides (organophosphate).

pigment: Artists' paints were in the past were extremely poisonous, but now are less of an issue. Examples: (a) Flake white is lead carbonate; (b) Cadmium red is cadmium sulphide; (c) Antimony white is antimony trioxide; (d) Barium yellow is barium chromate; (e) Chrome orange is lead carbonate; (f) Chrome yellow is lead chromate; (g) Cobalt violet is cobalt arsenate; (h) Naples yellow is lead antimonite; (i) Strontium yellow is strontium chromate; (j) Vermillion is mercury sulphide; (k) Zinc yellow is zinc chromate; and (l) Paris green is copper acetoarsenite.

poison, toxin, venom: Poison is a general term. Example: Arsenic added to someone's food or drink with intent to kill. A 'toxin' is a poisonous substance, or mixture of substances, usually of plant or animal origin, that can cause illness or death. Example: Mushrooms eaten that caused poisoning. A 'venom' is a poisonous substance or mixture of substances such as proteins and enzymes, produced by some animals to deter an aggressor or to attack prey. A venom intended to kill prey may contain enzymes that, once in the victim, begin digesting the victim's insides to turn it into a meal. An aggressor may be injured or killed. It is a poisonous fluid usually transmitted by bite or sting. Example: Snakes. It is thought that the Egyptian cobra, *Naja haje* was the asp in Cleopatra's suicide.

prunasin: A cyanogenic glycoside found in bitter almonds. It releases cyanide when eaten, as the body's digestive enzymes break the compound down.

prussic acid: see cyanide.

ptomaines: Compounds such as putrescine, cadaverine and neurine. The first two, probably the worst-smelling compounds known, are formed in putrefying flesh from the decomposition of the amino acids that were formed in the breakdown of proteins. Putrescine is from the amino acid arginine, and cadaverine is from the amino acid lysine. Neurine is formed when choline decomposes as flesh putrefies.

pulegone: see abortifacients.

putrescine: see ptomaines.

quinine: Poisonous alkaloid found in the bark of the cinchona tree. Used to treat malaria and add bitterness to tonic water.

radiation poisoning: Poisoning due to radiation is diverse and can be acute or chronic. There are three types of radiation: (a) Alpha radiation is due to high-energy positive particles, but has little capacity to penetrate materials. If a material emitting these particles enters the body, such as by inhalation of dust or ingestion of contaminated food, then the surrounding tissue is subjected to the radiation resulting in disease. (b) Beta radiation relates to high-energy negative particles and is more penetrating than the alpha radiation. It can pass through the skin and become absorbed by the body tissue where it damages cells. (c) The one with the most energy and with greatest penetrating power is gamma radiation, which is due to waves rather than particles. When body tissue is exposed to the gamma rays the high energy can damage or even destroy cells. Cells damaged by gamma rays may form cancers. Ironically gamma radiation is used in radiotherapy for killing cancer cells.

rodine: Rat poison in which the active ingredient is white phosphorus. As well as killing rats it will also poison humans if ingested.

ricin: Deadly poisonous protein found in the castor oil plant (*Ricinus communis*). One of the most poisonous chemicals known.

salts of lemon: Potassium oxalate. Was used for removing stains on linen. Causes poisoning in humans by producing calcium oxalate crystals in the kidneys.

scopolamine: see hyoscine.

Scoville scale: Some plants produce compounds that impart a burning sensation in the mouth of an animal exploring food sources. This is sufficient in most cases to deter the hungry animal. However, humans are perverse creatures and seem to take delight in the masochistic pleasure of food that hurts. The Scoville scale ranks the intensity of the mouth burn. An example of how effective this is occurs when chilli powder is mixed in with bird seed. The birds are unaffected and eat the seeds, which pass through their digestive system intact and are spread further afield in bird droppings. However, the squirrel hates the burning experience and is put off devouring the seeds, which would be destroyed in the squirrel's digestive tract. Thus, we see a plant poison working as a deterrent. At the top end of the scale is capsaicin, which is an irritant in small concentrations but a deadly poison in larger concentrations.

stibine: see antimony.

strychnine: Poisonous alkaloid from plant *Strychnos nux vomica.* Was used as a poison for rodents, a stimulant in tonics and an enhancer of athletic performance.

sugar of lead: see lead.

tetrodotoxin: A toxin found in puffer fish and other marine species. In the human body it reacts with nerve cells, prevents normal functioning and causes death. High concentrations are present in the ovaries and liver of puffer fish.

thallium: Thallium as a metal is similar to lead. Compounds, such as thallium sulphate are highly toxic. In the human body it mimics the role of potassium and can distribute itself throughout, but unlike potassium, it reacts destructively with the enzymes and destroys their essential functions with fatal results.

theobromine: Alkaloid found in tea, cocoa and chocolate, with dark chocolate having the greatest amount. Has similar effect in human body to that of caffeine. Poisonous to dogs.

toxication: Some substances are not directly poisonous but, due to their reactivity, they become toxic inside the human body. For example, methanol is regarded as poisonous as it can kill when ingested. It is not the methanol that causes death, but the formaldehyde and formic acid that is formed when the methanol reaches the liver and is broken down by the enzymes.

toxin: see poison.

volatile organic compounds: VOCs. Natural or man-made compounds that release significant amounts of vapour at room temperature. Indoors they degrade the quality of the air (see indoor air quality). Usually applies to man-made materials, but house plants also release some of these, whilst at the same time absorbing others.

venom: see poison.

vesicant: a chemical that causes blistering of the skin.

Bibliography

B. Bass and J. Jefferson, *Beyond the Body Farm*, William Morrow and Co.

G. Bell, *The Poison Principle*, Macmillan, London, Oxford and Basingstoke, 2002

J. Bennet and J. Conneally, *Crime Investigation*, Paragon Book Service Ltd.

J. C. Brenner, *Forensic Science Glossary*, CRC Press, Boca Raton, 2000.

D. G. Brown and E. V. Tullet, *Bernard Spilsbury: His Life and Cases*, George G. Harrap and Co Ltd, 1951.

The Merck Index, ed. S. Budavari, Merck & Co Inc, Rahway, NJ, 2001.

R. Carson, *Silent Spring*, Chapman and Hall, London, 1965.

J. Camp, *Magic, Myth and Medicine*, Priory Press Ltd, London, 1973.

J. D. Casswell, Q.C., *A Lance for Liberty*, George G. Harrap and Co Ltd, 1961.

T. Cook and A. Tattersal, *Senior Investigating Officers' Handbook*, OUP.

Criminal Statistics: England and Wales 2000, London, HMSO, Crown Copyright.

Introduction to Forensic Sciences, ed. W. G. Eckert, CRC Press, Boca Raton, 2nd edn, 1997.

J. J. Eddleston, *The Encyclopaedia of Executions*, John Blake, London, 2004.

Poisons and Poisonings: Death by Stealth
By Tony Hargreaves

Published by the Royal Society of Chemistry, www.rsc.org

S. Fielding, *The Hangman's Record, Volume One, 1868–1899*, Chancery House Press, Beckenham, Kent, 1994.

C. Emsley, *Crime and Society in England 1750–1900*, ed. Pearson, Harlow, 1996.

J. Emsley, *The Elements of Murder: A History of Poison*, Oxford University Press, 2005.

J. Emsley, *The Elements*, Oxford University Press, 1995.

J. Emsley, *The Consumer's Good Chemical Guide*, Corgi Books, 1996.

J. Emsley, *Nature's Building Blocks*.

J. Emsley, *Vanity, Vitality and Virility*, Oxford University Press, 2004.

J. Emsley and P. Fell, *Was it Something You Ate?*, Oxford University Press, 2002.

Z. Erzinclioglu, *Forensics: Crime Scene Investigation From Murder to Global Terrorism*, Carlton Books, London, 2000.

R. Evershed and N. Temple, *Sorting the Beef from the Bull*, Bloomsbury Sigma, ISBN 9781472911339.

J. Fraser and R. Williams, *Handbook of Forensic Science*, Willan Publishing, 2009.

R. Girling, Sunday Times, 28 November 2004, Cleopatra and the asp.

T. Hargreaves, Chemistry of death and decay, *Chem. Rev.*, November 2005, Philip Allan.

K. Inman and N. Rudin, *Principles and Practice of Criminalistics: The Profession of Forensic Science*, Protocols in Forensic Science Series, CRC, Boca Raton, 2001.

S. H. James and J. J. Nordby, *Forensic Science: Scientific and Investigative Techniques*.

J. Appl. Toxicol., Toxin Reviews, NY, M. Dekker.

Journal of the Forensic Science Society.

B. H. Kaye, *Science and the Detective*, VCH Publishers, New York, 1995.

J. Fraser, *Forensic Science: A Very Short Introduction*, Oxford University Press, 2010.

C. Fletcher, *Real Crime Scene Investigations*, Sumersdale Publishers, 2006.

Fundamentals of Applied Toxicology, Elsevier Science Ltd.

International Journal of Toxicology, Taylor and Francis.

Jackson and J. Jackson, *Forensic Science*, Pearson Prentice Hall, 2008.

A. Holden, *The St Albans Poisoner*, Panther, St Albans, 1976.

C. Kellet, *Poison and Poisoning*, Academic Press Ltd, 2012, ISBN 9781 90933 5059.

K. Othmer, *Encyclopaedia of Chemical Technology*, vol. 24.

B. Lane, *Encyclopaedia Forensic Science*, Magpie Books, London, 2004.

Standard Dictionary of Folklore, ed. M. Leach, Funk and Wagnalls.

T. A. Loomis and W. A. Hayes, *Loomis's Essentials of Toxicology*, Academic Press, 4th edn, 1996.

D. P. Lyle, *Forensics for Dummies*, Wiley Publishing Inc., 2004.

W. Marshall, The Bude Enigma Revisited, True Detective, November, 1997.

V. McDermid, *Forensics The Anatomy of Crime*, Profile Books Wellcome Collection, 2014.

Merck Index, Merck & Co. Inc., New Jersey, 1989.

H. Beers, *Merck Manual of Medical Information*, Pocket Books, New York, 2003.

S. E. Manahan, *Toxicological Chemistry*, Lewis Publishers, Boca Raton, 1992

T. Mueller, *Extra Virginity: The Sublime and Scandalous World of Olive Oil.*

T. Newburn, T. Williamson and A. Wright, *Handbook of Criminal Investigation*, Willan Publishing, 2007.

J. Nordley, *Forensic Science. Scientific and Investigation Techniques*, CRC Press.

Opie and Tatem, *Oxford Dictionary of Superstitions*, OUP.

Garner's Chemical Synonyms and Trade Names, ed. J. Pearce, Gower Technical Press, Aldershot, UK, 1987.

S. Rinpoche, *The Tibetan Book of Living and Dying*, Random House, 1992.

K. M. Ramsland, *Forensic science of CSI*, Berkley Publishing Group.

M. Roach, *Stiff: The Curious Lives of Human Cadavers*, Penguin Books, 2004.

N. Sly, *Murder by Poison, A Casebook of British Murders*, The History Press, 2009.

P. Stelfox, *Criminal Investigation: An Introduction to Principles and Practise*, Willan Publishing, 2009.

R. Saferstein, *Criminalistics: Introduction to Forensic Science*, Prentice Hall, 2003.

D. W. A. Sharp, *The Penguin Dictionary of Chemistry*, Penguin, London, 2003.

J. Simpson and S. Roud, *Oxford Dictionary of English Folklore*, OUP.

N. Sly, *Murder by Poison*.

D. Spoto, *The Biography of Marilyn Monroe*, Arrow Books Ltd, London, 1993.

O. Sterner, *Chemistry Health and Environment*, Wiley-VCH.

A. Summers, *Goddess: The Secret Lives of Marilyn Monroe*, Victor Gollancz Ltd, London, 1985.

J. Timbrell, *Introduction to Toxicology*, Taylor and Francis, 2002.

J. Timbrell, *The Poison Paradox: Chemicals as Friends and Foes*, Oxford University Press, 2005.

J. Trestrail, *Criminal Poisoning*, Humana Press, 2007.

J. Ward, *Crime Busting: Breakthroughs in Forensic Science*, Blandford Press, London, 1998.

K. Watson, *Poisoned Lives: English Prisoners and their Victims*, Hambledon Continuum, 2003.

P. C. White, *Crime Scene to Court*, RSC Publishing, 2004.

www.chemheritage.org

www.capitolpunishmentuk.org

www.emedicine.com

www.poison.org

www.planetdeadly.com

www.conservationinstitute.org

www.bbc.co.uk

www.h2g2.com

www.localhistories.org

www.rhs.org

www.independent.co.uk

www.sciencemag.org

www.world-nuclear.org

www.murderpedia.org

www.casebook.org

www.cbsnews.com

www.gutsandgore.com

Subject Index

Bold type refers to glossary entries. *Italic* refers to figures and illustrations.